Lactate in Acute Conditions

Lactate in
Acute Conditions

Editors
H. Bossart and C. Perret, Lausanne

60 figures and 38 tables, 1979

S. Karger · Basel · München · Paris · London · New York · Sydney

National Library of Medicine Cataloging in Publication
    International Symposium on Lactate in Acute Conditions, Basel, 1978
    Lactate in acute conditions/editors, H. Bossart and C. Perret
    Basel; New York: Karger, 1979
    1. Lactates – diagnostic use – congresses 2. Acute Disease – diagnosis – congresses
    I. Bossart, H., ed. II. Perret, Claude, ed. III. Title
    QU 98 I61L 1978
    ISBN 3-8055-2968-6

©   Copyright 1979 by S. Karger AG, 4011 Basel (Switzerland), Arnold-Böcklin-Strasse 25
Printed in Switzerland by Thür AG Offsetdruck, Pratteln
ISBN 3-8055-2968-6

Contents

Preface

Acute conditions are frequently linked with lack of oxygen, impaired perfusion of tissues and organs and therefore with lactate problems.

These emergency situations need quick diagnostic procedures. Unfortunately, clinical lactate determination did not meet this demand until a short time ago.

The new Roche electrochemical device gave lactate measurements a new practical interest, since results can be obtained within a few minutes.

This symposium was therefore organized to answer the following questions:

(1) What do we now know about lactate in acute conditions? (2) Is lactate really representative for hypoxic conditions and, if so, in which ones? (3) Can the 'bedside measurement' of lactate improve clinical handling of the acute patient? (4) Do we have any reason, in the light of this new technique, to do further work and possibly organize another symposium on lactate within the next few years?

We hope that this work may contribute to a better understanding of some acute problems in the critically ill patient.

C. PERRET
H. BOSSART

Basic Aspects

Lactate in Acute Conditions. Int. Symp., Basel 1978, pp. 1–9 (Karger, Basel 1979)

Cell Metabolism and Lactate

W. Schumer

Department of Surgery, University of Health Sciences, The Chicago Medical School at Veterans Administration Hospital, North Chicago, Ill.

Introduction

Lactate was first identified as an acidic constituent of foods by SCHEELE [12] in 1780, and has been industrially manufactured by fermentation since 1881. The stereoisomeric form has been named D- and L-lactate referring to its molecular structural relationship with the glyceric acid isomer, as shown in figure 1. D(−) lactate, or dextro, has been considered a non-physiologic isomer ever since CORI's [5] classical experiment showed that D(−) lactate is poorly metabolized, and that 30–40% of the lactate ingested is excreted in the urine. L(+) lactate is considered to be the physiologic isomer. However, recent investigations have shown that D(−) lactate can be metabolized by the rat, and probably the human, thus placing the definitive value of the lactate determinations in question. Are we measuring lactate with biochemical methods that are specific of L(+) lactate? Since most of our present clinical testing is specific of L(+) lactate, then we must assume that we may not be measuring all the lactate in the human serum.

Lactate's pK of 3.86 is less than the pH range of body fluid, and it dissociates freely. The ratio of lactate ion-lactic acid in the body pH range is 3,000 to one. To study the cellular metabolism of lactate one must refer to the classic Cori cycle: L(+) lactate is produced as a result of muscle cell anaerobiosis, and/or the ordinary glycolytic oxidation of glucose in all cells. In non-stressed, or non-shocked animals enough lactate is produced to maintain a concentration of 0.7–1 mM/l. It has recently been estimated that lactate is produced in the resting man at the following rates (mM/h/kg): skeletal mass, 3.13; brain, 0.14; red cell mass, 0.18; and 0.11 for blood elements, renal medulla, intestinal mucosa and skin. Total lactate production in a 70-kg man is approximately

1

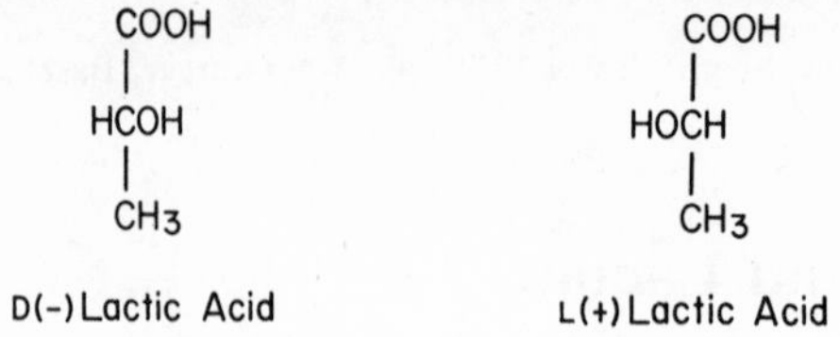

COOH
|
HCOH
|
CH$_3$

D(-)Lactic Acid

COOH
|
HOCH
|
CH$_3$

L(+)Lactic Acid

2

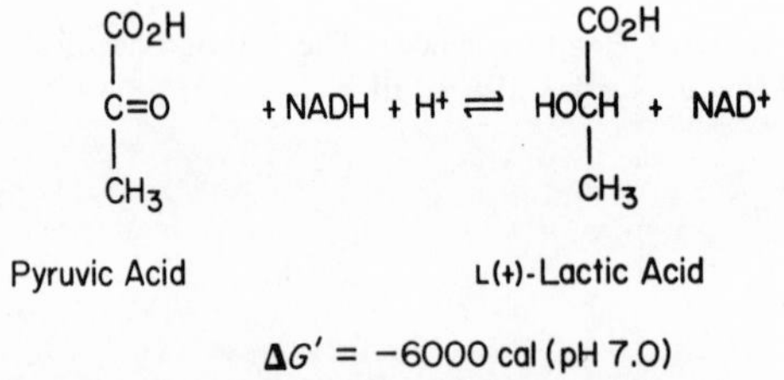

CO$_2$H
|
C=O   + NADH + H$^+$ $\rightleftharpoons$ HOCH   +   NAD$^+$
|
CH$_3$

Pyruvic Acid

CO$_2$H
|
HOCH
|
CH$_3$

L(+)-Lactic Acid

$\Delta G' = -6000$ cal (pH 7.0)

Fig. 1. Stereoisomeric form of lactate (D- and L-lactate); molecular structural relationship with the glyceric acid isomers.

Fig. 2. Conversion of lactate to pyruvate via the NAD-dependent lactate dehydrogenase reaction.

1,300 mM/day. Approximately 53% of this load is metabolized by the liver [1]. In the liver cell, lactate is freely diffused through the hepatocyte membrane and immediately converted to pyruvate via the NAD-dependent lactate dehydrogenase reaction, as shown in figure 2.

Liver Lactate Metabolism

Since the lactate dehydrogenase reaction is the entrance into the gluconeogenic scheme, pyruvate, because of thermodynamic inhibitions is converted to oxaloacetate, as shown in figure 3. This reaction is aerobically catalyzed by pyruvate carboxylase, and proceeds from oxaloacetate to phosphoenolpyruvate. The rate-limiting steps in lactate's conversion to glucose occur at the pyruvate carboxylase reaction, the phosphoenolpyruvate to 2-phosphoglycerate reaction, and at the conversion of fructose-1, 6-phosphate to fructose-6-phosphate. Our recent studies in endotoxic and septic shock models have shown that gluconeogenic inhibition of the liver cell markedly increases cellular and serum lactate [14]. This inhibition, occurring at the rate-limiting steps in the gluconeogenic scheme, renders the liver and kidney unable to utilize the predominant non-carbohydrate precursor of glucose which is lactate. Thus, lactate which may be pro-

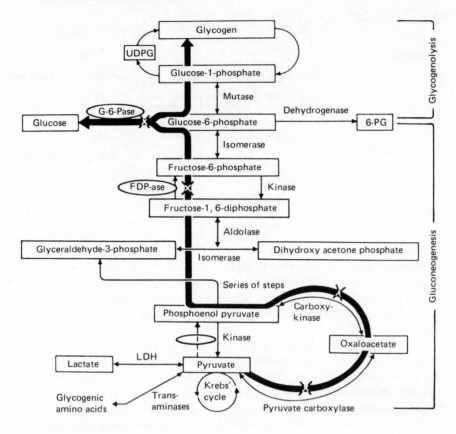

Fig. 3. Conversion of lactate to pyruvate and oxaloacetate in the gluconeogenic scheme.

duced in excessive quantities because of muscle cell anaerobiosis will not be metabolized.

Lactate is utilized in the liver and kidney and produced in the muscle, as a result of lactate dehydrogenase. Lactate dehydrogenase isoenzymes consist of two types of polypeptide chains in five possible combinations; M_4, M_3H, H_2M_2, MH_3, and H_4. The M_4 and M_3H isoenzymes are predominantly found in tissues highly dependent on glycolysis for energy, such as skeletal muscle; and the MH_3 and H_4 isoenzymes predominate in tissues with an aerobic or respiratory metabolism. Kinetic studies of the properties of the different lactate dehydrogenase isoenzymes have revealed that those from rapidly glycolysing tissues, such as muscle, have a very high affinity for pyruvate as an electron acceptor, and those from actively respiring tissues have a relatively low affinity [6]. The lactate

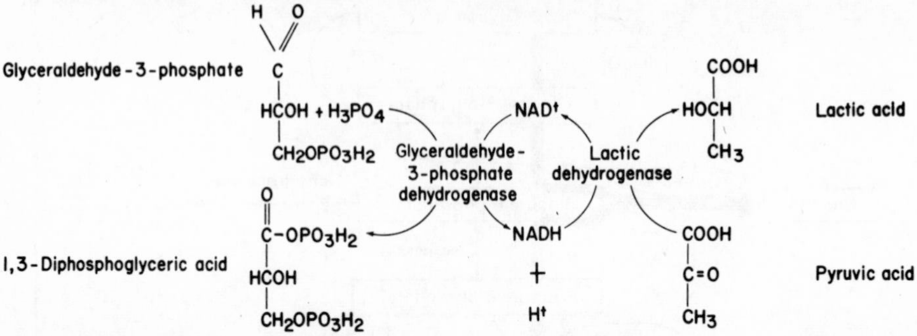

Fig. 4. Anaerobic oxidation of NADH.

dehydrogenase isoenzymes present in the muscle cell effectively reoxidize reduced nicotine adenine dinucleotide (NADH) with pyruvate to produce lactate; while liver isoenzymes are less reactive with pyruvate and allow NADH to be more readily reoxidized aerobically by the mitochondria. These isoenzymes, therefore, control the direction of the lactate-pyruvate reaction.

Another factor influencing the direction of this reaction is NADH availability. If NADH is plentiful because of the donated electrons from the glyceraldehyde-3-phosphate to 3-phosphoglycerate reaction, then the reaction equilibrium is directed to the right as demonstrated by the large negative value of $\Delta G'(-6.0 \text{ kcal})$ as shown in figure 4. This is what occurs in the shock state with its attendant increased anaerobiosis and electron transport system's inability to oxidize NADH [10].

The other factor influencing the augmented production of lactate is glycolysis, a thermodynamically feasible, naturally occurring process. Contrariwise, gluconeogenesis functions against thermodynamic principles and, therefore, needs energy in the form of adenosine triphosphate (ATP) to function. An energy deficit will inhibit gluconeogenesis, and then lactate will accumulate in the cytosol and in the serum [3].

Muscle Lactate Metabolism

Muscle cell glycolysis involves two stages: the first stage, needing two ATPs, is the collection of simple sugars and the conversion of glyceraldehyde phosphate. The second stage is the oxidoreduction of glyceraldehyde to lactate with a coupled formation of four ATPs. Figure 5 illustrates these two stages. Lactate, the end product of the glycolytic

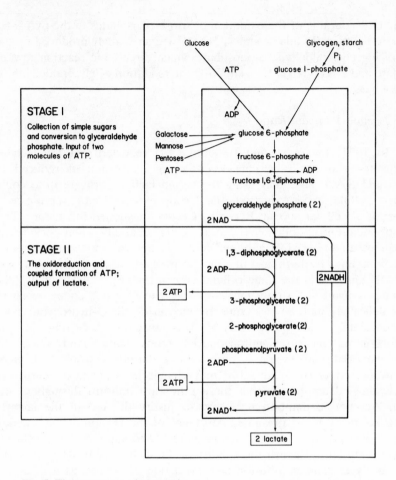

Fig. 5. The two stages of glycolysis.

sequence under anaerobic conditions, will diffuse through the cellular plasma membrane into the plasma. During short bursts of vigorous activity by the muscle cell, lactate escapes into the blood in large quantities. Muscle fiber fatigue or rigor is partly due to lactate acidification.

Since anaerobic metabolism must function uninterruptedly in anoxic states, then the oxidation of glyceraldehyde-3-phosphate is crucial to the glycolytic cell. As shown in figure 3, oxidized nicotine adenine dinucleotide (NAD) is the primary oxidizing agent accepting electrons in the oxidation of triose-phosphates. The amount of NAD being limited, the reaction will deteriorate as soon as all the NAD is reduced. There is a symbiotic coupling of the oxidation of triose-phosphates to the pyruvate-lactate reaction. In anoxic states, NADH is increasingly produced by the oxidizing glyceraldehyde-3-phosphate which directs the reaction toward lactate and NAD. NAD then supports the oxidation of glyceraldehyde-3-phosphate.

Lactate-Pyruvate Ratio

In 1941, FRIEDMAN and BARBORKA [7] reported that the lactate-pyruvate ratio could be considered as an expression of reduced to unbound NAD. And that using the serum lactate-pyruvate ratio could reflect cellular oxidoreduction, the lactate-pyruvate ratio serving as a monitor of cellular anoxia. Recently, COHEN questioned this theory. This ratio's monitoring ability loses some of its reliability when considering that most of the NADH is bound and may not participate in the redox reaction, and furthermore the lactate-pyruvate ratio and NAD and NADH must be in the same cellular compartment [4]. Additionally, the lactate-pyruvate ratio only measures the cytosol redox state; however, this deficiency may be overcome by measuring the 3-hydroxybutyrate-acetoacetate ratio (OHB-AcAc). This reaction occurring in the mitochondria measures mitochondrial redox state. And since the mitochondrial redox system is more reduced than the cytosol, the lactate-pyruvate and OHB-AcAc ratios may change in opposite directions. A misleading assumption is that these ratios are uniform throughout the cell. Another assumption necessary to justify the use of the lactate-pyruvate ratio, is the presumed constancy of the H^+ ion at the reaction site. However, there are no studies available to support this constancy. Further militating against the validity of the lactate-pyruvate ratio is the variability in cellular diffusion and extraction of lactate, as well as the variability of the redox state in the different organs. Taking into consideration the deficit in all these assumptions, the blood redox ratios remain as the only means available to measure tissue anoxia.

Recent investigations have proposed a biochemical basis for the treatment of lactic acidosis [17]. As mentioned above, the only biochemical reaction that produces or consumes lactate is the interconversion of lactate and pyruvate, or the lactate dehydrogenase reaction. Lactate metabolism depends upon two major pathways for removal of pyruvate: (1) the pyruvate dehydrogenase reaction leading to pyruvate oxidation via the Krebs' cycle or pathway, occurring in all cells, and (2) the pyruvate carboxylase reaction toward gluconeogenesis from pyruvate, occurring mainly in the liver and kidney cortex.

Lactic acidosis occurs whenever the lactate production rate exceeds

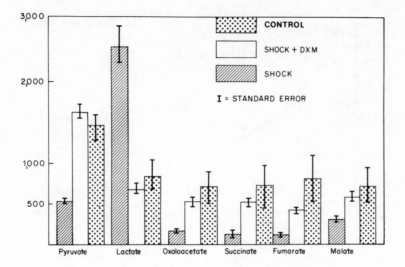

Fig. 6. Dexamethasone phosphate (DXM) effect on the glycolytic and Krebs' cycle intermediates of liver in shock.

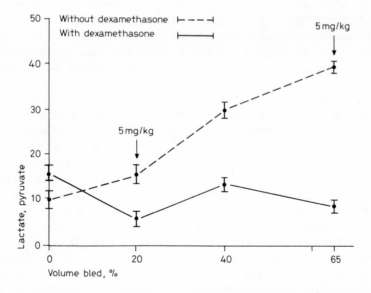

Fig. 7. Dexamethasone effect on the lactate-pyruvate ratio in oligemic shock.

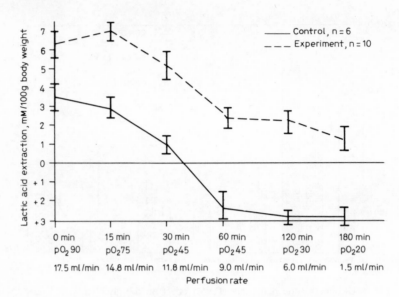

Fig. 8. Glucagon effect on perfusant lactate in the *in vitro* liver perfusion.

the utilization rate. Many hormones, such as glucocorticoids and glucagon, affect the gluconeogenic rate-limiting reactions by inducing pyruvate carboxylase synthesis and stimulating lactate metabolism via gluconeogenesis. Our studies have indicated that lactic acidosis in septic, endotoxic, and hemorrhagic shock can be alleviated by the administration of pharmacologic doses of steroids and physiologic doses of glucagon, as shown in figures 6–8 [9, 14, 16].

Recent investigations of a simple substance, dichloroacetate, have shown that it affects lactate metabolism by activating pyruvate dehydrogenase, the enzyme responsible for initiating pyruvate oxidation and consequently decreasing lactate levels [11]. STACPOOLE *et al.* showed, clinically, dichloroacetate's salutary effect on lactate reduction. Although his patients were diabetics with minimally elevated lactate levels, other investigators have reported dichloroacetate's marked effectiveness in decreasing blood lactate in dogs poisoned with biguanides [11, 17].

It is essential to study the biochemistry of lactate because of its crucial effect in the clinical setting. In 1966, we reported lactate's role in the production of irreversibility in low flow states in the dog. The simple administration of either lactic acid or 0.1 normal hydrochloric acid caused metabolic acidemia and microcirculatory alterations characteristic of low flow states [15]. These results emphasized the need for continued studies on cellular lactate metabolism, specifically, the development of a therapeutic mechanism to counteract the deleterious effect of excessive serum or blood lactate.

References

1 COHEN, P. J.: Editorial views. More on lactate. Anesthesiology 43: 614–616 (1975).
2 COHEN, R. D.: Disorders of lactic acid metabolism IV; in BESSER, BIERICH, BONDY, DAUGHADAY, FRANCHIMONT and HALL Clinics in endocrinology and metabolism, vol. 5, pp. 613–625 (Saunders, Philadelphia 1976).
3 COHEN, R. D.; ILES, R. A.; BARNETT, D.; HOWELL, M. E. O., and STRUNIN, J.: The effect of changes in lactate uptake on the intracellular pH of the perfused rat liver. Clin. Sci. 41: 159–170 (1971).
4 COHEN, R. D. and SIMPSON, R.: Lactate metabolism. Anesthesiology 43: 661–673 (1975).
5 CORI, C. F.: Mammalian carbohydrate metabolism. Physiol. Rev. 11: 143–275 (1931).
6 EVERSE, J.; BERGER, R. L., and KAPLAN, N. O.: Physiological concentrations of lactate dehydrogenases and substrate inhibition. Science 168: 1236–1238 (1970).
7 FRIEDMANN, T. E. and BARBORKA, C.: The significance of the ratio of lactic to pyruvic acid in blood after excercise. J. biol. Chem. 141: 993–994 (1941).
8 GIESECKE, D. and FABRITIUS, A.: Oxidation and excretion of D-lactic acid by rats. Experientia 30: 1124 (1974).
9 HOLTZMAN, S.; SCHULER, J. J.; EARNEST, W.; ERVE, P. R., and SCHUMER, W.: Carbohydrate metabolism in endotoxemia. Circ. Shock 1: 99–105 (1974).
10 KEUL, J.; DOLL, E., and KEPPLER, D.: Oxidative energy supply IV; in JOKL and HEBBELINCK Energy metabolism of human muscle, Med. and Sport, vol. 7, pp. 105–136 (University Park Press, Baltimore 1972).
11 RELMAN, A. S.: Editorial. Lactic acidosis and a possible new treatment. New Engl. J. Med. 298: 564–565 (1978).
12 SCHEELE, K.: In M. Borillon LaGrange. 30. Germinal An. XII. An. Chem. Phys. 40: 270 (1780).
13 SCHRÖDER, R.; ELTRINGHAM, W. K.; GUMPERT, J. R. W.; JENNY, M. E.; PLUTH, J. R., and ZOLLINGER, R. M., jr.: The role of the liver in the development of lactic acidosis in low flow states. Post-grad. med. J. 45: 566–570 (1969).
14 SCHULER, J. J.; ERVE, P. R., and SCHUMER, W.: Glucocorticoid effect on hepatic carbohydrate metabolism in the endotoxin-shocked monkey. Ann. Surg. 183: 345–354 (1976).
15 SCHUMER, W.: Lactic acid as a factor in the production of irreversibility in oligohaemic shock. Nature, Lond. 212: 1210–1212 (1966).
16 SCHUMER, W.; MILLER, B.; NICHOLS, R. L.; McDONALD, G. O., and NYHUS, L. M.: Metabolic and microcirculatory effects of glucagon in hypovolemic shock. Archs Surg., Lond. 107: 176–180 (1973).
17 STACPOOLE, P. W.; MOORE, G. W., and KORNHAUSER, D. M.: Metabolic effects of dichloroacetate in patients with diabetes mellitus and hyperlipoproteinemia. New Engl. J. Med. 298: 526–530 (1978).

W. SCHUMER, MD, Department of Surgery, University of Health Sciences, The Chicago Medical School, Veterans Administration Hospital, North Chicago, IL 60064 (USA)

Lactate in Acute Conditions. Int. Symp., Basel 1978, pp. 10–19 (Karger, Basel 1979)

The Production and Removal of Lactate

R. D. COHEN

Metabolic and Endocrine Unit, The London Hospital, London

Any consideration of the pathogenesis of a disorder of lactate metabolism always raises the question: Is too much being produced, or is lactate removal failing? In this talk, I shall firstly review some aspects of the detailed mechanisms by which tissues release lactate into the circulation or take it up and then give a more physiological overview of whole body lactate homeostasis. I am going to pay special attention to present areas of uncertainty or ignorance, since one of the main purposes of a conference such as this is to stimulate further work.

The two detailed aspects I am going to discuss are, firstly, mechanisms of translocation of lactate across cell membranes and secondly the interactions of acid-base disturbances and lactate production and removal. Remarkably little attention has been paid in the past to the first of these topics. Lactate has often been described as 'freely diffusible'. To me, this means 'just like urea'. But the simple fact that the lactate ion is charged immediately precludes so simplistic a view. The first question is whether lactate crosses cell membranes in the ionized form L^- or as the undissociated HL. We then have to consider whether translocation is passive or active. Passive transport under the influence of the electrochemical or concentration gradients can be by simple movement through membrane channels in the case of L^- or by lipid solubility – so-called non-ionic diffusion – for HL. In addition, some carrier-assisted translocations are also passive.

We have previously reviewed [1] the evidence provided by many workers that lactate can move across cell membranes in either form – certainly in muscle and liver – and I only propose to give two examples taken from our own studies on perfused rat liver. The first is from some old work in which we reasoned that if significant quantities of lactate entered the liver cell as L^-, the pH of the hepatocyte should rise with increasing lactate uptake. If entry was on the other hand entirely as HL, no change in cell pH should occur. The equations below give the overall equations of lactate conversion to glucose – either from HL or L^-.

$$2CH_3CHOHCOOH \rightarrow C_6H_{12}O_6,$$

$$2CH_3CHOHCOO^- + 2H_2O \rightarrow C_6H_{12}O_6 + 2OH^-.$$

Only if L^- gets in does a hydroxyl ion appear on the right hand side of the equation. The same applies if lactate ions are converted to CO_2 and water rather than glucose. Experimentally, we find that pH_i does rise with increasing lactate uptake [2, 3]; so some of the lactate must get in in the ionized form. Much more recently, we have examined in rat perfused liver the relationship between the hydrogen ion concentration ratio across the cell membrane (H_i/H_e) and the lactate concentration ratio (L_e/L_i). At low lactate concentration (approximately 1 mmol/l), there is a significant relationship, and with a slope of 0.77, indicating substantial entry as HL. At both moderate (2.5 mmol/l) and higher ($\simeq 4$ mmol/l) lactate concentrations, there is no relationship at all, suggesting that transport is mainly in the ionized form. We now have to ask the question whether or not there has to be a pump to get lactate across the hepatocyte membrane. The answer is fairly straight forward so far as the lactate ion is concerned; the electrochemical gradient for L^- can be calculated from the membrane potential, which is about 40 mV, and the lactate concentration ratio, and it turns out that L^- would have to be pumped to get into the cell at all.

The situation is a little more difficult for HL. Can HL enter cells by simple non-ionic diffusion through the membrane fast enough to account for the observed lactate uptake or even a substantial part of it? To answer this question, one has to calculate the concentration of HL each side of the membrane and have a figure for the permeability coefficient of HL across the membrane. To get an idea of the permeability coefficient for HL, we have compared it [ILES and COHEN, unpublished] in an artificial system with the weak acid 5,5-dimethyl oxazolidine 2,4-dione (DMO) for which we already have measured the permeability coefficient for the liver cell membrane. The artificial system consists of an aqueous phase, separated by Visking membrane from a lipid phase. The aqueous phase has a pH <1 so that both ^{14}C-DMO and ^3H-lactate are virtually completely un-ionized. In the lipid phase, we have placed a large variety of lipid solvents of varying polarity, and measured the simultaneous rate of penetration of HDMO and HL. HL has always entered the lipid phase very much more slowly than HDMO. We have therefore assumed that the liver cell membrane is much less permeable on the basis of lipid solubility alone to HL than to HDMO. Now even if we use the HDMO permeability coefficient for HL, the rate of simple non-ionic diffusion of HL can be shown to be 2–3 orders of magnitude less than the observed rate of uptake of lactate. We therefore suggested [4] that if HL forms any substantial part of the lactate uptake, it must also be pumped, or there must be some form of facilitated diffusion.

It is therefore of some considerable interest that two fairly specific lactate transporting systems have been identified in red cell membranes by HALESTRAP [5] and in the membrane of the Ehrlich ascites tumour cell by SPENCER and LEHNINGER [6]. The first system is the general anion transporter and the second is a carboxylate transporter. Both are inhibited by cyanocinnamate compounds and only the first by 4,4-di-isothiocyanostilbene 2,2-disulphonic acid (DIDS). There is no real indication as yet as to whether these transporters are parts of energy-requiring pumps or whether they simply facilitate passive transport. But the clinical importance of such pumps and transporters is that inhibition of such mechanisms might be potent causes of disorders such as lactic acidosis. For instance, much work has been done on the effect of phenformin on the biochemical mechanisms of gluconeogenesis, but no one to my knowledge has investigated whether phenformin inhibits a lactate transporter in the cell membrane, or the somewhat similar transporter of pyruvate into the mitochondrial membrane. In this context, SCHÄFER [7, 8] has shown dramatic effects of biguanides on the structure and permeability of mitochondrial and artificial phospholipid membranes.

I would now like to turn to the second topic, namely the interactions of acid-base status and lactate production and removal. Looking at production first of all, it has been found in all tissues tested that glycolysis is inhibited by acidosis and stimulated by alkalosis. Analysis of intermediate metabolite concentrations indicate that the phosphofructokinase (PFK) step is the site where changes in pH exert their effect, and this conforms to expectations from studies of PFK *in vitro*, where inhibition by ATP is strikingly modified by pH changes. There are some interesting physiological consequences of this effect. For example, in skeletal muscle the accumulation of lactate and hydrogen ions during anaerobic exercise cause a striking fall in cell pH which eventually inhibits glycolysis. This may be one of the mechanisms of fatigue, and thus prevents completely unlimited lactic acid production, which would otherwise by overwhelming the capacity of the removal mechanisms produce a degree of acidosis incompatible with life. This negative feedback system also produces an internal partial buffering system for the muscle cell – the production of lactic acid by lowering pH_i prevents further H^+ production (Fig. 1b).

If we now turn from organs which produce lactate to the liver, we have a very different situation. Unlike in muscle, here we have a positive

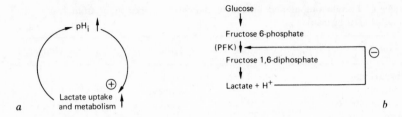

Fig. 1. Interrelation of cell pH (pH$_i$) and lactate metabolism (a) in liver and (b) muscle, brain and erythrocytes. PFK = Phosphofructokinase. The encircled plus and minus indicate the nature of the feedback. See text for further discussion.

feedback system. The left hand limb in figure 1(a) represents the effect of lactate uptake and metabolism in causing alkalinization of the cell pH which I talked about earlier. On the right hand, we have another effect which is demonstrable in the perfused liver, namely that the rise in cell pH (achieved by lowering PCO$_2$) further stimulates gluconeogenesis from lactate and speeds lactate uptake. One can see that this positive feedback system is teleologically well placed to deal with increases in lactate levels in blood reaching the liver, for instance after exercise. Unfortunately, if lactate uptake is inhibited, the positive feedback works the other way and lactate uptake is rapidly shut off. We have suggested [4, 9] that this may be an important factor in the development of lactic acidosis.

We have shown by freeze clamping acidotic perfused livers and measuring intermediate metabolites, including oxaloacetate, that the point where acid cell pH inhibits gluconeogenesis from lactate is between pyruvate and oxaloacetate [10]. This could be a direct effect of intracellular acidosis on the activation of pyruvate carboxylase by acetyl CoA, but could also have other explanations, such as acidosis lowering the cytosolic oxaloacetate concentration by altering redox ratios within the cell.

The other main organ of lactate uptake, namely the kidney, behaves rather differently to liver because acidosis stimulates rather than depresses lactate uptake. The reason for acute stimulation [11] is unknown, but does not involve stimulation of gluconeogenesis. More chronic stimulation depends on the appearance of increased activity of PEPCK in the kidney [12]; this is a rate-limiting enzyme in gluconeogenesis both in the liver and the kidney, but unlike in the liver, the kidney enzyme is stimulated by acidosis.

We must now consider the whole body situation, where the first problem is to determine the rate of entry of lactate to the lactate pool and the rate of exit, under various conditions – for instance rest, exercise,

Table I. Lactate output and uptake of individual tissues in or from resting man (mol/24 h/70-kg man)

Tissue	Output	Tissue	Uptake
Erythrocytes	0.30	Liver	0.72
Brain	0.22	Kidneys	0.12
Skeletal muscle	0.20	Heart	0.08
Skin	0.37	Skeletal muscle	?
Leucocytes	0.03		
Platelets	0.03		
Renal medulla	0.02		
Intestinal mucosae	0.10		
Total output	1.27	Total uptake	0.92+

This table is modified from COHEN and WOODS [9], in which the sources of the individual data are given, with the exception of that from the kidney, which is taken from NIETH and SCHOLLMEYER [25].

anoxia. Conceptually, the easiest approach is to estimate directly by measuring arteriovenous lactate differences and flow rates, the output and uptake of various organs and sum them. Though conceptually simple, this method is highly invasive, and data for man is rather fragmentary. Table I shows the results of collating evidence from the literature in resting man [9]. The units are mol/24 h/70 kg man. As the Fick principle cannot be applied to red and white cells, skin, renal medulla and intestinal mucosa, values have had to be inserted which have been derived from incubation *in vitro* under simulated physiological conditions. For skeletal muscle, the values have been proportioned up from forearm studies to the whole body. Adding these up, we have a resting output of 1.27 mol per day. On the uptake side, we can only in man account for two thirds of the output. I shall return later to the difficult problem of whether skeletal muscle takes up lactate under resting conditions. I would like to point out that with this 1.27 mol of lactate are also produced 1.27 mol of H^+ which are eliminated when lactate is taken up by the removal organs. This amount of H^+ is an order of magnitude greater than the kidneys excrete daily, a fact which seems to be ignored by many physiologists.

Because of the difficulties in the approach taken in this slide, many workers have taken advantage of isotopic dilution techniques to measure lactate turnover [13, 14]. The commonest protocol consists of a primed constant infusion of ^{14}C-lactate.

When blood specific activity (SA) and lactate concentrations are constant, turnover is calculated by dividing the rate of infusion of counts by the blood lactate SA. The model which this calculation must presuppose consists of simple entry of counts into a single compartment and one way exit of lactate, now labelled. But of course the model is really far more complicated than this. The simplest elaboration of the model possible adds an assumed single pyruvate pool, so that now counts can return from the pyruvate pool into the lactate pool. Only if the carbon flow through the pyruvate pool from non-lactate sources is infinite compared with that from lactate could the first model be identified with the second. Another approach is to assume that lactate and pyruvate specific activity very quickly equalize. The two pools could then be regarded as one and the measured lactate turnover might under some circumstances approximate the total carbon flow through the lactate and pyruvate pool. However, measurement of the SA of both pyruvate and lactate is seldom undertaken, so the validity of this assumption has not been established; it would be expected that the error introduced by it would vary from one set of conditions to another. Estimates of lactate turnover in man using this technique are some 20–60% in excess of those using summation of individual organ contributions previously discussed [13, 14]. In dogs, the high blood lactate of haemorrhage shock has, using the isotope dilution technique, been variously attributed to either predominant overproduction or predominant under-removal of lactate [15–17]. The scanty evidence in man which has been obtained by measurement of a-v differences shows that in cardiogenic or endotoxic shock the liver and kidneys produce rather than remove lactate in half the cases studied [18]. The conclusion therefore seems unavoidable that deletion of removal mechanisms must be a major factor in the hyperlactaemia as well as overproduction.

A more ambitious isotopic approach is to determine the net fate of lactate by measuring the appearance of label in glucose and CO_2 during primed constant infusion of labelled lactate [14]. Thus, it is hoped to measure the rate of gluconeogenesis from lactate and rate of complete oxidation of lactate. But here again, the isotopic approach leads to serious interpretational problems. The major difficulty is that the model used for the calculations assumes that the oxaloacetate molecules derived from the labelled lactate and which are destined for either oxidation or gluconeogenesis are not allowed to mix with molecules derived from other substrates for oxidation and gluconeogenesis, and which have no labelling. This of course is not the case in practice, for oxidation and gluconeogenesis share a common oxaloacetate pool in the mitochondria. The consequence of the true situation is that, for instance, the gluconeogenic pathway cannot distinguish oxaloacetate molecules according to whether they were derived from lactate or other *unlabelled* substrates and the specific activity of the resulting glucose is lower than it should be; the percentage of glucose derived from lactate is therefore underestimated. Similar troubles occur with the calculation of oxidation rates from labelled CO_2. KREBS *et al.* [19] showed in kidney slices incubated with a mixture of acetoacetate and labelled lactate that, although acetoacetate does not form glucose and accounted for most of the fuel of respiration, the specific activity of the respiratory CO_2 was much greater than could be accounted for by the very small contribution of lactate to respiration and the specific activity of glucose formed was much less than would have been expected from that of the labelled lactate. KREBS *et al.* concluded that where two metabolic pathways 'crossover', i.e. in this case at oxaloacetate, it is impossible by considering the fate of the label derived from a substrate to draw valid conclusions about the net fate of that substrate. Rather similar arguments have been advanced by STEELE *et al.* [20].

Since the errors implicit in isotope dilution techniques are undetermined and may presumably vary from one situation to another, it is my view that significant advances in our understanding of disorders of lactate metabolism in man will come not from isotope dilution studies but from careful measurement of arteriovenous differences and flow rates across individual organs, for example the liver. KREISBERG *et al.* [14] point out the difficulties of measuring organ substrate balance in man and emphasise the importance of validating the relatively non-invasive isotope techniques against the reference standard. But this has to my knowledge not yet been systematically achieved in disorders of lactate metabolism. The difficulties to which KREISBERG *et al.* refer – of measuring a-v differences and flow rates – are to a large extent ethical in nature. For instance, I would be personally unhappy about delaying treatment with bicarbonate in phenformin-induced lactic acidosis in order to place a hepatic venous catheter. But I believe that useful information can still be obtained *after* the start of treatment, and this information might turn out to be beneficial to the individual patient himself as well as to future patients with lactic acidosis. Of course, such investigation should only be carried out at centres which for other reasons are expert at vascular catheterization procedures. This whole issue obviously needs to be carefully debated.

Another approach to determining how the body handles lactate is to administer a lactate load intravenously and to determine its fate. I would like to discuss two pieces of work in this context. Firstly in our own laboratory, Dr. YUDKIN and I [21] have attempted to determine the contribution of the kidney to the removal of a lactic acid load by comparing the disappearance curves in conscious normal and nephrectomized rats. Under normal acid-base conditions, the kidney seems responsible for removal of about 20% of the load. If this is done on rats made previously acidotic to various degrees by feeding ammonium chloride, then the contribution of the kidney rises to about 40% as the arterial pH falls to below 7.0. Thus, the kidney compensates for a decrease in the contribution of the rest of the body, so that ability of the whole body to remove lactic acid remains remarkably constant under these conditions; but in shock or when poisoned with phenformin, the kidney may not be able to adjust in this way. In man, there is some very careful new work by CONNOR et al. [22] which shows how the removal of a sodium lactate load is slowed in liver disease.

This brings me to the relative roles of liver and skeletal muscle in lactate removal. Table I showed that in resting man approximately 50–60% of the total lactate removal might be accomplished by the liver. Other direct estimates of hepatic lactate uptake in the literature might give values somewhat lower, but in particular I would like to draw your attention to the work of AHLBORG et al. [23] who infused sodium lactate into resting volunteers and concluded from direct measurements of regional uptake that skeletal muscle metabolized 35% and the splanchnic area 10% of the lactate infused. They suggested that skeletal muscle is quantitatively more important than the liver in the disposal of lactate in the resting state. However, their data show that the sodium lactate infusion resulted in a 9% expansion of extracellular space and a 2.5-fold increase in blood flow in the leg and 35% increase in O_2 uptake. Now skeletal muscle is in *the resting state* a tissue with a poor blood supply, bordering on anaerobiosis, and it would not be surprising if the increased blood supply due to the volume expansion permitted increased uptake of lactate – and other substrates – for oxidation. But in normal man and in disease when an endogenous lactate load is added to the circulation, it is added as (lactate $+ H^+$), not (lactate $+ Na^+$). (Lactate $+ H^+$) causes no osmotic expansion of extracellular fluid since the H^+ destroys HCO_3^-, which is merely replaced by lactate. Thus, the conclusions from these studies about the proportion of lactate removed by skeletal muscle may be seriously affected by artefacts arising from experimental procedures which involve sodium lactate rather than lactic acid infusion. In addition to this criticism, there is some uncertainty as to how much skeletal muscle uptake merely represents filling of the lactate space rather than actual metabolism. I must emphasise that although there are little doubts about the major role of skeletal muscle in removing lactate in some circumstances during submaximal exercise [24], the average patient developing lactic acidosis is not indulging in exercise.

It is difficult to come to any conclusion other than that there are major uncertainties in our quantitative understanding of whole body lactate turnover in health and disease and of the relative roles of the different organs in production and removal. This may sound depressing but I believe it is time to take a critical look at the present state of knowledge, and I hope that some of the points made above will contribute to this review.

Summary

More attention needs to be paid to the mechanisms of translocation of lactate across cell membranes. The multiple interactions between lactate metabolism and acid-base status may play an important role in lactate physiology both in health and disease. The relative contributions of the different organs to lactate uptake and removal are still poorly understood in quantitative terms. Definitive advances in the understanding of lactate pathophysiology are likely to come from direct organ balance studies rather than isotope dilution techniques.

References

1   COHEN, R. D. and ILES, R. A.: Lactic acidosis – some physiological and clinical considerations. Clin. Sci. mol. Med. *53:* 405–410 (1977).
2   COHEN, R. D.: ILES, R. A.; BARNETT, D.; HOWELL, M. E. O., and STRUNIN, J.: The effect of changes in lactate uptake on the intracellular pH of the perfused rat liver. Clin. Sci. *41:* 159–170 (1971).
3   LLOYD, M. H.; ILES, R. A.; SIMPSON, B. R.; STRUNIN, J. M.; LAYTON, J. M., and COHEN, R. D.: The effect of simulated metabolic acidosis on intracellular pH and lactate metabolism in the isolated perfused rat liver. Clin. Sci. mol. Med. *45:* 543–549 (1973).
4   COHEN, R. D. and ILES, R. A.: Intracellular pH: Measurement, control and metabolic interrelationships. CRC Crit. Rev. clin. Lab. Sci. *6:* 101–143 (1975).
5   HALESTRAP, A. P.: Transport of pyruvate and lactate into human erythrocytes: evidence for the involvement of the chloride carrier and a chloride independent carrier. Biochem. J. *156:* 193–207 (1976).
6   SPENCER, T. L. and LEHNINGER, A. L.: L-Lactate transport in Ehrlich ascites tumour cells. Biochem. J. *154:* 405–414 (1976).
7   SCHÄFER, G.: Some new aspects of the interaction of hypoglycaemia-producing biguanides with biological membranes. Biochem. Pharmacol. *25:* 2015–2029 (1976).
8   SCHÄFER, G.: On the mechanism of action of hypoglycaemia-producing biguanides. A reevaluation and a molecular theory. Biochem. Pharmacol. *25:* 2005–2014 (1976).
9   COHEN, R. D. and WOODS, H. F.: Clinical and biochemical aspects of lactic acidosis, pp. 137–145 (Blackwell, Oxford 1976).
10 ILES, R. A.; COHEN, R. D.; RIST, A. H., and BARON, P. G.: The mechanism of inhibition by acidosis of gluconeogenesis from lactate in rat liver. Biochem. J. *164:* 185–191 (1977).

11 YUDKIN, J. and COHEN, R. D.: The effect of acidosis on lactate removed by the perfused rat kidney. Clin. Sci. mol. Med. *50:* 177–184 (1976).

12 ALLEYNE, G. A. O. and SCULLARD, G. H.: Renal metabolic response to acid-base changes. I. Enzymatic control of ammoniagenesis. J. clin. Invest. *48:* 364–370 (1969).

13 SEARLE, G. L. and CAVALIERI, R. R.: Determination of lactate kinetics in the human; analysis of data from single injection vs. continuous infusion methods. Proc. Soc. exp. Biol. Med. *139:* 1002–1006 (1972).

14 KREISBERG, R. A.; PENNINGTON, L. F., and BOSCHELL, B. R.: Lactate turnover and gluconeogenesis in normal and obese humans. Diabetes *19:* 53–63 (1970).

15 WIENER, R. and SPITZER, J. J.: Lactate metabolism following severe haemorrhage in the conscious dog. Am. J. Physiol. *227:* 58–62 (1973).

16 ELDRIDGE, F. L.: Relationship between lactate turnover and blood concentration in haemorrhagic shock. J. appl. Physiol. *37:* 321–323 (1974).

17 DANIEL, A. M.; PIERCE, C. H.; MCLEAN, L. D., and SHIZGAL, H. M: Lactate metabolism in the dog during shock from haemorrhage, cardiac tamponade or endotoxin. Surgery Gynec. Obstet. *143:* 581–586 (1976).

18 SRIUSSADAPORN, S. and COHN, J. N.: Regional lactate metabolism in clinical and experimental shock. Circulation *37:* suppl. 6, p. 187 (1968).

19 KREBS, H. A.; HEMS, R.; WEIDEMANN, M. J., and SPEAKE, R. N.: The fate of isotopic carbon in kidney cortex synthesizing glucose from lactate. Biochem. J. *101:* 242–249 (1966).

20 STEELE, R; WINKLER, B.; RATHGEB, I.; BJERKNES, C., and ALTSZULER, N.: Plasma glucose and free fatty acid metabolism in normal and long-fasted dogs. Am. J. Physiol. *214:* 313–319 (1968).

21 YUDKIN, J. and COHEN, R. D.: The contribution of the kidney to the removal of a lactic acid load under normal and acidotic conditions in the conscious rat. Clin. Sci. mol. Med. *48:* 121–131 (1975).

22 CONNOR, H.; WOODS, H. F.; MURRAY, J. D., and LEDINGHAM, J. G. G.: The kinetics of elimination of a sodium-lactate load in man: the effect of liver disease. Clin. Sci. mol. Med. *54:* 338–348 (1978).

23 AHLBORG, G.; HAGENFELDT, L., and WAHREN, J.: Influence of lactate infusion on glucose and FFA metabolism in man. Scand. J. clin. Lab. Invest. *36:* 193–201 (1976).

24 HERMANSEN, L. and STENSVOLD, I.: Production and removal of lactate during exercise. Acta physiol. scand. *86:* 191–201 (1972).

25 NIETH, H. and SCHOLLMEYER, P.: Substrate utilization of the human kidney. Nature, Lond. *209:* 1224–1225 (1966).

R. D. COHEN, MD, Metabolic and Endocrine Unit, The London Hospital, Whitechapel Road, *London E1 1BB* (England)

Lactate in Acute Conditions. Int. Symp., Basel 1978, pp. 20–28 (Karger, Basel 1979)

Methodology of Lactate Assay

I. KRAGENINGS

Institut für Klinische Chemie am Klinikum Grosshadern der Universität München, München

The time elapsed between the initial attempts to derive a method for the determination of a substance and the first final, widely accepted result, is often an indication of the importance of the substance. In the case of lactate, the first attempts were made over 70 years ago.

In the view of the clinical chemist, the methodological development of lactate determination is a typical example for progress in this discipline as a whole; early rough gravimetric or titrimetric, time-consuming methods that required large amounts of sample, then a colorimetric method combined with troublesome reagents and unpleasant treatment were introduced. They gave way to exact enzymatic analyses in an enzyme-linked system with inert reagents, which are quick, highly specific and sensitive.

Colorimetric methods are today mostly of historical interest but, nevertheless, many results of importance for the clinicians were acquired with these methods. The basis of all colorimetric methods was the conversion of lactate by means of heat and concentrated sulfuric acid into acetic aldehyde, carbon monoxide and water. The acetic aldehyde thus formed reacted with a phenol derivate to form a colour complex, whose intensity could be determined spectrophotometrically against a standard or from a calibration curve. The best colorimetric method was described by BARKER and SUMMERSON [1] in 1941, in which the acetic aldehyde is determined by its colour reaction with p-hydroxydiphenyl in presence of cupric ions. According to the literature, this method must have given sufficient results, and the latest modification of this method was published in 1972 [8].

New technologies to prepare and purify enzymes brought new aspects. Improvement of methods to isolate the specific enzyme lactic acid dehydrogenase (LDH), the availability of commercial preparations and finally the test kits made the enzymatic reaction the method of choice.

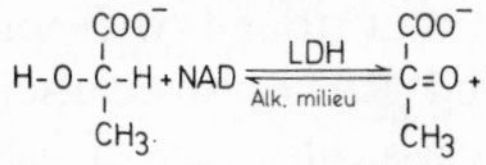

$$\underset{\overset{|}{CH_3}}{\overset{\overset{COO^-}{|}}{H-O-C-H}} + NAD \underset{Alk.\ milieu}{\overset{LDH}{\rightleftharpoons}} \underset{\overset{|}{CH_3}}{\overset{\overset{COO^-}{|}}{C=O}} + NADH + H^+$$

Fig. 1. Principles of measuring lactate 1: LDH–NAD system. Measurement: Spectrophotometrical-direct registration in UV spectrum of the formed NADH. Spectrophotometrical-indirect registration in visible spectrum of the NADH complex coupled to an oxidoreduction dye. Origin of LDH: Animal tissue from heart, muscle or stomach.

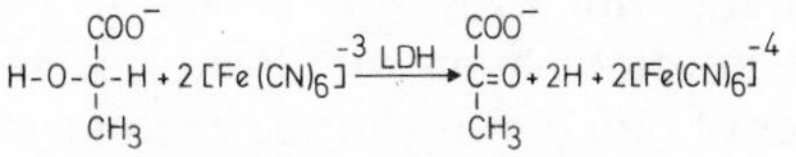

$$\underset{\overset{|}{CH_3}}{\overset{\overset{COO^-}{|}}{H-O-C-H}} + 2\,[Fe\,(CN)_6]^{-3} \overset{LDH}{\longrightarrow} \underset{\overset{|}{CH_3}}{\overset{\overset{COO^-}{|}}{C=O}} + 2H + 2[Fe(CN)_6]^{-4}$$

Fig. 2. Principles of measuring lactate 2: LDH–Fe (III) complex system. Measurement: Spectrophotometrical registration of the absorbance in reduction, the Fe (III) complex. Amperometric registration of the current necessary to re-oxidize the reduced Fe (III) complex. Origin of LDH: Yeast.

The introduction of the specific enzyme LDH resulted in substantially two different methods:

Group 1, combination of muscle LDH and NAD (fig. 1): (1.1) with spectrophotometric registration of the absorbance of the formed NADH in the UV spectrum [5, 7, 9, 11, 14, 15, 25, 26]; (1.2) with spectrophotometric registration in the visible spectrum by coupling the NADH to an oxidoreduction dye [4].

Group 2, combination of yeast LDH with a ferric iron complex (fig. 2): (2.1) with spectrophotometric registration of colour after the reduction [29]; (2.2) with amperometric registration of the current needed in re-oxidation of the iron (III) complex [6, 18, 23].

(1.1) The methods in the first group have been in use longer and seemed to be more successful. The enzyme is obtained here from different types of muscle tissue from various animals. Catalyzed by the enzyme, the lactate is oxidized to pyruvate. The liberated electrons are transferred to NAD (nicotineadeninedinucleotide) to form NADH, which in reduced form is capable of absorbing light in the UV spectrum, the increase in absorbance being directly proportional to lactate concentration. Fluorimetric measurement was also tried [20]. The equilibrium of this reaction is totally shifted to the left. By alkaline ambiance and eliminating the formed pyruvate, the reaction is forced to the side of pyruvate. The elimination of pyruvate is achieved by destroying the pyruvate with peroxide [27] by trapping with a carbamyl group as semicarbazon [15, 24] or hydrazine [11, 15, 25] or by a subsequent enzymatic reaction with, or transamination by, the enzyme glutamic pyruvic transaminase (GPT) [19].

A variation of this method involves the use of 3-acetyl analog APAD instead of NAD [16], but this variation has not found widespread application. The combination of muscle LDH and NAD proved itself to be the most reliable method and became the 'classical' assay of lactate determination. The name of HOHORST [11] is inevitably mentioned in this context. Attempts to improve this method included only minor changes. Different buffers were tried: TEA [7, 16, 20], Tris-HCl [28], carbonate [19, 27], arsenite [5], though glycine buffer seemed to be the most suitable [11, 25, 27].

Phosphate buffer in combination with hydrazine results in higher blank values, while arsenite buffer in combination with hydrazine gave a more stable blank value. A number of different hydrazine concentrations were tested, enzyme concentrations were varied depending on their origin.

(1.2) For photometric registration in the visible part of the spectrum the reduced NAD is bound by the enzyme diaphorase to an oxidoreduction dye. The dye INT (3-p-nitrophenyl-2-iodophenyl-tetrazolium-chloride) [4] has commonly been used for this purpose as it has a narrow absorption peak at 500 nm. This type of reaction is most suitable for automatic analysis, particularly in a continuous flow system [10].

(2) The reduction of iron (III) complexes by absorption of hydrogen ions during the oxidation of lactate to pyruvate is the basis of measurement in a NAD-independent system. The enzyme in this reaction is obtained from yeast.

(2.1) This reaction can be registered photometrically in the visible range of the spectrum, hexacyanoferrate was used predominantly as the electron acceptor but iron (III)-phenanthroline [30] is equivalent. In a phosphate buffer environment with a pH of 8, the equilibrium of the reaction is totally weighted to the right. The method is simple, but the need to isolate the enzyme from yeast prevented its widespread use.

(2.2) The introduction of an enzymatic electrochemical sensor brought an improvement in the method. Here too, as previously mentioned, the NAD-independent enzyme LDH from yeast (here called cytochrome b_2) is catalyzing the oxidation of lactate to pyruvate in the presence of hexacyanoferrate (III) as an electron acceptor. By a platinum electrode biased against a silver-silverchloride electrode, the formed hexacyanoferrate (II) produced in the reaction is re-oxidized and the current needed for the re-oxidation is registered with a galvanometer. The catalyzing enzyme is fixed in a layer on the platinum electrode, which is separated from the sample chamber by a semipermeable membrane (fig. 1). This fixed enzyme is not discarded after each measurement. The catalyzing properties of LDH can be used for many measurements. The principle of this type of measurement is realized in the Lactate Analyzer 640 produced by Hoffmann-La Roche.

In order to discuss fully all the possibilities of lactate determination, one should mention an instrument recently introduced by Eppendorf that is not yet readily available on the market: the MLF 6620. This instrument also operates with a coated enzyme. Lactate is oxidized enzymatically to acetic aldehyde, the oxygen consumption is measured via an O_2 electrode and hence the lactate concentration is calculated.

The methods outlined according to their chemical processes will now be described in more detail. The reaction catalyzed by yeast LDH and using iron (III) complex as electron acceptor has not found widespread use. Durliat et al. [3] did use it, however, in a comparative study of different enzyme electrodes and found it equivalent to others.

The following discussion, therefore, will substantially be a comparison of modifications of reactions catalyzed by muscle LDH and NAD and enzymatic-electrochemical reactions. To begin with, the sample distribution will be taken as the start of each analysis. Most investigators work with protein-free samples when using methods based on the reaction of NAD-dependent LDH. Whole blood or plasma is deproteinized in different ratios. The most suitable agent for protein precipitation seems to be trichloric acid in concentration between 3 and 6%. Marbach and Weil [17] used 5% metaphosphoric acid with good results. A substantial advantage of protein precipitation is that the sample thus treated is stable over a long period of time. Hochella and Weinhouse [10] are the only ones who use serum in a continuous flow system. They admit, though, that the sample suffers from a lack of stability and that it has to be dealt with very quickly. Plasma is used in conjunction with a glycolysis inhibitor (usually NaF) only in discrete analyzers in which the sample is first very much diluted. Such systems are the fast analyzer [21] and DuPont ACA [28] which operate with prepared reagent packs. A recently introduced Mono-Test (Boehringer) is making it possible to use plasma directly with a reagent kit. In contrast, if measuring with an enzymatic-electrochemical sensor, plasma or serum and even whole blood can be investigated directly. In our experience, values obtained from whole blood are not as easily reproducible as those obtained from plasma or serum. The dilution of the sample (1:10) can easily be carried out with a dilutor. Plasma with NaF or oxalate is not suited for estimation in the lactate analyzer, because the enzyme is cumulatively inhibited by these substances.

In the classic enzymatic-spectrophotometric test, sample collection and determination are more time-consuming than in the lactate analyzer. Discounting the times required for sample collection, transport to the laboratory and centrifugation, it is only possible to get the results in approximately 7 min in a fast analyzer or in the DuPont ACA; by the Mono-Test, the value is obtained within 20 min. But this depends on the availability of a suitable apparatus and of a skilled technician. Estimating with the lactate analyzer, the result can be used within 3–5 min for further clinical decisions. A definite advantage of the enzyme electrode lies in its flexibility and in the possibility of bedside measurement.

Analytical methods in clinical chemistry are described by their performance characteristics: linearity, accuracy, precision, specificity and correlation to related methods. The criteria presented are results from a study together with Dr. RACKWITZ, and Prof. JAHRMÄRKER from the Medical Clinic [13].

The several points of reliability were:

Linearity. The upper limit of the enzymatic-spectrophotometric methods was 15 mmol/l. For the manual method of HOHORST, we found 15 mmol/l, for the modification for a discrete analyzer, the Braun Systematik, 16 mmol/l, and for the DuPont ACA 17 mmol/l. The limits should not be taken as absolute values as they depend on the reagents and on the measuring system used. Using the Lactate Analyzer 640, the linearity extended to a value of 30 mmol/l, while the manufacturer specified a value of 20 mmol/l.

Accuracy. The recovery of added lactate was 96–103% both for enzymatic-spectrophotometric methods as well as for the Lactate Analyzer 640. Testing the control sera with assigned values for lactate, we found good agreement between the true value and our results.

Precision. The coefficient of variation for precision within run varies from 1.4 to 3.27%, between day precision from 2.71 to 5.8% for the

Table I. Precision in lactate measurement

Method	x̄	s	n	CV
Precision within run				
Enzymatic-spectrophotometric				
Manual [HOHORST]	4.33	0.142	20	3.27
DuPont ACA	4.50	0.052	20	1.14
Braun Systematik	3.32	0.094	20	2.83
Enzymatic-amperometric				
Lactate Analyzer 640	4.0	0.05	20	1.25
Precision day-to-day				
Enzymatic-spectrophotometric				
Manual [HOHORST]	4.37	0.163	20	3.73
DuPont ACA	4.17	0.113	10	2.71
Braun Systematik	3.35	0.194	20	5.8
Enzymatic-amperometric				
Lactate Analyzer 640	4.00	0.126	20	3.15

Table II. Interfering substances

Enzymatic-spectrophotometric	Enzymatic-amperometric
α-Hydroxy-butyrate	1, 3-phenyllactic acid
β-Chlor-lactate	glyceric acid
Glycerine	2-hydroxy-isocaproic acid
Glycol acid	2-hydroxy-hexanoic acid
α-Glycerophosphate	2-hydroxy-butyric acid
	uric acid at high concentrations

different enzymatic-spectrophotometric methods (table I). The precision within run for the lactate analyzer was 1.25% and between day precision was 3.15%. These coefficients of variation corresponded to values found in other routine methods of clinical chemistry.

Specificity. This means the ability of an analytical method to determine solely the components it is intended to measure. The specificity of the various methods for the determination of lactate is good [6], though not absolute. Possible disturbing influences are given in table II. None of these substances is found in blood in such concentrations that they interfere severely.

The correlation between the individual methods is good, as shown in the figures 3–5. The coefficients of correlation were r = 0.969 and r = 0.994 (fig. 3–5); no significant differences were revealed by the Wilcoxon test. This finding agrees well with those of other investigators [2, 12, 22].

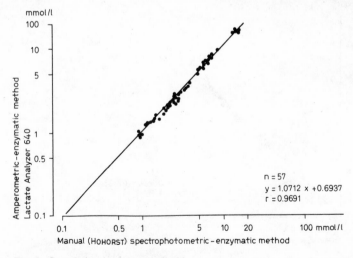

Fig. 3. Correlation between methods.

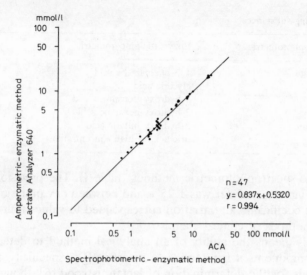

Fig. 4. Correlation between methods.

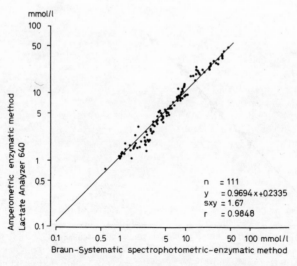

Fig. 5. Correlation between methods.

Summary

Two different methods for the determination of lactate in biological fluids are employed in the clinical chemist's laboratory. Both are based on the enzymatic conversion of lactate to pyruvate:

(1) Conversion of lactate by the NAD-dependent enzyme LDH from animal muscle and direct photometric registration of the formed NADH (reduced form of NAD) in UV spectrum, or registration in visible spectrum after coupling to an oxidoreduction dye.

(2) Conversion of lactate by the NAD-independent LDH (cytochrome b_2) from yeast, transferring the freed electrons to hexacyanoferrate and registration of the current necessary to re-oxidize the hexacyanoferrate.

The methods of the first group are more time-consuming and mostly require a protein-free filtrate of the material. They have the advantage of being adaptable to all mechanized systems and are preferable, especially for all scientific purposes with a large number of samples. The second one, the enzymatic-electrochemical method, has the advantage of a flexible, small, easy-to-handle apparatus. Plasma or serum are preferable to whole blood. The different types of methods are equal regarding precision, accuracy and specificity. The correlation between the methods is good, the results obtained with different methods are well comparable.

References

1   BARKER, S. B. and SUMMERSON, W. H.: The colorimetric determination of lactic acid in biological material. J. biol. Chem. *147:* 415 (1941).

2   DITESHEIM, P. J. et BOSSART, H.: Premiers essais de mesures du L-lactate plasmatique au moyen du 'Lactate Analyzer 5400'. Schweiz. med. Wschr. *106:* 1598–1601 (1976).

3   DURLIAT, H.; COMTAT, M., and BAUDRAS, A.: Spectrophotometric and electrochemical determinations of L(+)-lactate in blood by use of lactate dehydrogenase from yeast. Clin. Chem. *22:* 1802–1802 (1976).

4   FRIEDLAND, I. M. and DIETRICH, L. S. : A rapid enzymic determination of L(+)-lactic acid. Analyt. Biochem. *2:* 390–392 (1961).

5   GERCKEN, G.: Die quantitative enzymatische Dehydrierung von L(+)-Lactat für die Mikroanalyse. Hoppe-Seyler's Z. physiol. Chem. *320:* 180–186 (1960).

6   GUARNACCIA, R. and RACINE, P.: Investigation on the specificity of an amperometric enzymatic sensor L-lactate. Biomed. Technik *21:* suppl., pp. 189–190 (1976).

7   HADJIVASSILIOV, A. G. and RIEDER, S. V.: The enzymatic assay of pyruvic and lactic acids. A definitive procedure. Clinica chim. Acta *19:* 357–361 (1968).

8   HARROWER, J. R. and BROWN, C. H.: Blood lactic acid – a micromethod adapted to field collection of microliter samples. J. appl. Physiol. *32:* 709–711 (1972).

9   HESS, B.: Über eine kinetisch-enzymatische Bestimmung der L(+) Milchsäure im menschlichen Serum und anderen biologischen Flüssigkeiten. Biochem. Z. *328:* 110 (1956).

10 HOCHELLA, N. J. and WEINHOUSE, S.: Automated lactic acid determination in serum and tissue extracts. Analyt. Biochem. *10:* 301–317 (1965).

11 HOHORST, H. J.: in BERGMEYER Methoden der enzymatischen Analyse; 2. Aufl., vol. II, p. 1425 (Verlag Chemie, Weinheim 1970).

12 KLENK, H. O.; GUARNACCIA, R., and MINDT, W.: Stat assay of lactate – a new clinial tool (Hoffman-La Roche, Basel 1976).

13 KRAGENINGS, I. und RACKWITZ, R.: Bestimmung von Laktat nach enzymatisch-elektrochemischem Prinzip im Vergleich mit drei Modifikationen der enzymatischen Methode. Lactate determination by the enzymatic-electrochemical principle in comparison to three modifications at the enzymatic method. Ärztl. Lab. *23:* 549–554 (1977).

14 LUNDHOLM, L.; MOHME-LUNDHOLM, E., and SWEDMYR, N.: Comparative investigation of methods for determinations of lactic acid in blood and in tissue extracts. Scand. J. clin. Lab. Invest. *15:* 311–316 (1963).

15 LUNDHOLM, L.; MOHME-LUNDHOLM, E., and VAMOS, N.: lactic acid with L(+) lactic acid dehydrogenase from rabbit muscle. Acta physiol. scand. *1963:* 243–249 (1963).

16 MAURER, C. and POPPENDIEK, B.: in BERGMEYER Methoden der enzymatischen Analyse, vol. II, pp. 1518–1521 (Verlag Chemie, Weinheim 1974).

17 MARBACH, E. P. and WEIL, M. H.: Rapid enzymatic measurement of blood lactate and pyruvic. Klin. Chem. *13:* 314–324 (1967).

18 MINDT, W.; RACINE, P. und SCHLÄPFER, P.: Sensoren für Lactat und Glucose. Ber. Bunsenges. phys. Chem. *77:* 805–808 (1973).

19 NOLL, F.: in BERGMEYER Methoden der enzymatischen Analyse; 3. Aufl., vol. II, p. 1521 (Verlag Chemie, Weinheim 1974).

20 PASSONNEAV, J. V.: in BERGMEYER Methoden der enzymatischen Analyse, vol. III, pp. 1515–1518 (Verlag Chemie, Weinheim 1974).

21 PESCE, M. A.; BODOURIAN, S., and NICHOLSON, J. F.: Rapid kinetic measurement of lactate in plasma with a centrifugal analyzer. Clin. Chem. *21:* 1932–1934 (1975).

22 RACINE, P.; KLENK, W. D., and KOCHSIEK, K.: Rapid lactate determination with an electrochemical enzymatic sensor: clinical usability and comparative measurements. Z. klin. Chem. klin Biochem. *13:* 533–539 (1975).

23 RACINE, P. and MINDT, W.: On the role of substrate diffusion in enzyme electrodes. Biol. Aspects Electrochem. *18:* suppl., pp. 525–534 (1971).

24 ROSENBERG, J. C. and RUSH, B. F.: An enzymatic-spectrophotometric determination of pyruvic and lactic acid in blood. Clin. Chem. *12:* 299–307 (1966).

25 SCHOLZ, R.; SCHMITZ, H.; BÜCHER, T. und LAMPEN, J. O.: Über die Wirkung von Nystatin auf Bäckerhefe. Biochem. Z. *331:* 71–86 (1959).

26 TFELT-HENSEN, P. and SIGAARD-ANDERSEN, O.: Lactate and pyruvate determination in $50\mu l$ whole blood. Clinica chim. Acta *19:* 357 (1968).

27 WARBURG, O.; GAWEHN, K. und GEISLER, A. W.: Weiterentwicklung der zell-physiologischen Methoden. Verbindung von Manometrie und optischer Milchsäure-Bestimmung. Hoppe-Seyler's Z. physiol. Chem. *320:* 277–279 (1960).

28 WESTGARD, J. O.; LAHMEYER, B. L., and BIRNBAUM, M. L.: Use of the Du Pont 'Automatic Clinical Analyzer' in direct determination of lactic acid in plasma stabilized with sodium fluoride. Clin. Chem. *18:* 1334–1338 (1972).

29 WIELAND, O.: Eine optisch-enzymatische Bestimmung der L(+) Milchsäure mit DPN-unabhängiger Milchsäuredehydrogenase aus Hefe. Biochem. Z. *329:* 568–576 (1958).

30 WIELAND, O. und JAGOW-WESTERMANN, B.: Bestimmung mit Lactat-Dehydrogenase aus Hefe; in BERGMEYER Methoden der enzymatischen Analyse, vol. II, p. 1442 (Verlag Chemie, Weinheim 1970).

I. KRAGENINGS, MD, Institut für Klinische Chemie am Klinikum Grosshadern der Universität München, Marchioninistrasse 15, *D-8000 München 70* (FRG)

Lactate in Acute Conditions. Int. Symp., Basel 1978, pp. 29–47 (Karger, Basel 1979)

Maternal and Fetal Lactate Characteristics during Labor and Delivery[1]

J. A. Low, S. R. Pancham, W. N. Piercy, D. Worthington and J. Karchmar[2]

Department of Obstetrics and Gynaecology, Queen's University at Kingston, Kingston, Ont.

Introduction

The interpretation of maternal and fetal metabolic acidosis during the intrapartum period requires an appreciation of the changes which occur in respect to lactate and pyruvate in the wide spectrum of 'normal' and 'complicated' pregnancies encountered in obstetric practice. It is the objective of this report to consider three questions on the basis of maternal and fetal observations in 911 pregnancies: (1) the changes which occur in maternal and fetal lactate concentration during labor and delivery, (2) the mechanisms which account for these changes, and (3) the management of lactate and pyruvate by the placenta.

The concentration of lactate and pyruvate in the mother and fetus is increased during labor and delivery [10, 16, 23, 24, 26, 28, 29, 33, 37]. However, the studies reported in the literature differ in regard to the characteristics of the patients assessed and the method, site and scheduling of samples obtained. These differences must be noted when specific measures are being compared.

It is widely recognized that an increase of lactate concentration may be due to an increase of pyruvate alone, or an increase of pyruvate and tissue hypoxia acting simultaneously [20] in keeping with the equation:

$$\text{Lactate} = \text{pyruvate} \times K \frac{\text{DPNH}_2}{\text{DPN}}.$$

Thus, an increase of lactate may be due to an increase of pyruvate with a parallel rise of lactate with no change of the lactate-pyruvate ratio, or due

[1] Supported by Ministry of Health Grant 473.

[2] The assistance of Mr. L. Broekhoven, Department of Mathematics, Queen's University, with the statistical analysis of the data is gratefully acknowledged.

to alteration of the oxygenation of the diphosphonucleotide (DPN) of the lactic dehydrogenase system resulting in an increase of lactate independent of pyruvate with a corresponding increase of the lactate-pyruvate ratio.

An understanding of the role of the placenta in respect to the transfer and metabolism of lactate and pyruvate is essential to the interpretation of maternal and particularly fetal lactate and pyruvate estimations. Animal studies with one exception [12] indicate that fetal lactate does not usually cross the placenta to the mother [5, 21, 22, 25, 36, 38].

Methods

This study includes maternal and fetal observations from 911 pregnancies. There were 57 patients with uncomplicated pregnancies delivered by elective section; 146 patients with 'normal' pregnancies, i.e. no maternal medical or obstetrical complications, who following a normal labor delivered a normal mature fetus; and 708 patients with 'complicated' pregnancies, i.e. who exhibited one or more maternal medical, obstetrical, gestational, labor or delivery complication. The complications observed in these patients are summarized in table I. Maternal medical complications included hypertension defined as essential hypertension or chronic renal disease with a blood pressure greater than 140/90, and diabetes including both diabetes predating the present pregnancy and gestational diabetes. Toxemia was defined as a systolic pressure greater than 140 mm Hg and a diastolic pressure greater than 90 mm Hg, and/or significant albuminuria. Antepartum hemorrhage in the second half of pregnancy included abnormal bleeding due to placenta previa, premature placental separation or an unknown cause. Abnormal labor included abnormal uterine action defined as uterine contractions with abnormal features in respect to frequency, intensity and duration associated with delayed cervical dilatation with or without dystocia confirmed by radiological examination. The congenital anomalies included only 'major' anomalies. Intrauterine growth retardation was defined as the fetus less than the 10th percentile by weight for gestational age.

The observations made include maternal venous blood acid-base, glucose, lactate and pyruvate characteristics during labor and at delivery, fetal acid-base characteristics during the last half of labor, and acid-base, glucose, lactate and pyruvate characteristics at delivery.

Measurements of pH were made with a micro electrode pH meter (AME-1 Astrup). Oxygen tension and carbon dioxide tension were measured with an IL 133-S1 system with O_2 (Clark) and CO_2 (Severinhaus) electrodes. Buffer base was calculated from the Singer-Hastings nomogram with the pH and pCO_2 and corrected for the degree of oxygen unsaturation. Blood sugar was measured by the glucose oxidase method.

Protein precipitation with 10% trichloroacetic acid was carried out within 3 min of sampling for the measurement of lactate and pyruvate and the filtrates were stored in an ice water bath. Lactate determinations were carried out in duplicate by the method of Barker and Summerson [2]. Pyruvate determinations were carried out by the partial extraction method of Friedman and Haughan as modified by Huckabee [18]. The blood water content was determined, and results of lactate and pyruvate concentration determinations were expressed as millimoles per liter of blood water.

The significance of the association between the factors relating to increased production of lactate and pyruvate have been assessed by a multiple regression analysis and significance is expressed in terms of probability (p) values.

Table I. The maternal, obstetrical, fetal and labor complications occurring in the patients of the 'complicated' pregnancy group

Maternal medical complications		Labor complications	
Hypertension	35	Abnormal labor	248
Diabetes	43	Delivery complications	
Cardiac disease	8	Mid forceps	201
Others	58	Other	31
Obstetrical complications		Section	112
Toxemia	160	Fetal complications	
Antepartum hemorrhage	42	Preterm	98
Multiple pregnancy	17	Post-term	30
Clinical fetus distress		Congenital anomaly	16
Meconium	191	IUGR	103
Meconium with abnormal FHR	83		

Table II. The acid-base and glucose observations in maternal and fetal blood during labor and at delivery

	Maternal				Fetal					
	labor			delivery	labor, time prior to delivery				delivery	
	onset	mid	2nd stage		mid	2 h	1 h	15 min	umb. vein	umb. art.
	N-428	N-277	N-330	N-911	N-224	N-123	N-198	N-239	N-910	N-885
pH	7.401	7.416	7.411	7.385	7.304	7.281	7.276	7.265	7.318	7.245
BB, m/Eq/l	42.0	41.9	41.9	41.0	43.0	42.3	41.6	41.1	41.4	39.5
pCO_2, mm Hg	32.0	30.0	31.0	32.0	47.0	51.0	50.0	51.0	39.7	50.8
pO_2, mm Hg	48.0	48.0	47.0	57.0	19.1	18.6	17.9	17.5	26.8	15.6
Glucose, mg%	106	111	103	130					119	100

Results

Measurements of maternal and fetal acid-base and glucose characteristics during labor and delivery are outlined in table II. There is a small decrease of buffer base in maternal venous blood at delivery with a corresponding decrease of pH. The fetus similarly exhibits a small decrease of buffer base during the last half of labor and at delivery with a decrease of pH. Maternal glucose concentration increased during labor

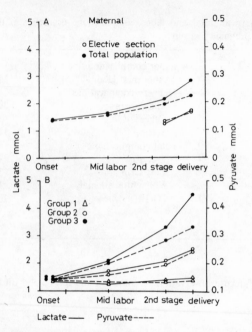

Fig. 1. Maternal lactate and pyruvate concentrations. *A* Maternal values during labor and at delivery for the total population and prior to and at delivery for the elective section group. *B* Maternal values during labor and at delivery for the three maternal lactate groups.

Table III. The increase of lactate and pyruvate occurring in maternal venous blood during the period of labor and the period of delivery

	Labor	Delivery	Total
Lactate, mmol	0.75	0.65	1.40
Pyruvate, mmol	0.05	0.04	0.09

and particularly at delivery. Fetal glucose concentration at delivery as measured in the umbilical vein and artery is also increased in relation to observations reported in the normal fetus at the onset of labor [27].

Maternal and Fetal Lactate and Pyruvate Concentrations during Labor and Delivery

The maternal venous blood lactate at the onset of labor was 1.4 mmol and pyruvate 0.14 mmol with an L/P ratio of 10. The increase during labor and delivery was 1.4 mmol for lactate and 0.09 mmol for pyruvate,

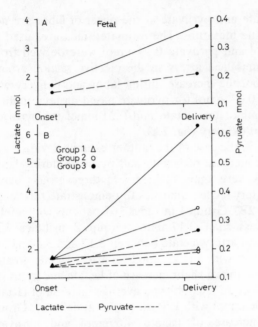

Fig. 2. Fetal lactate and pyruvate concentrations. *A* Projected increase of fetal values during labor and delivery for the total population. *B* Projected increase of fetal values during labor and delivery for the three fetal lactate groups.

so that at delivery, maternal lactate was 2.8 mmol and pyruvate 0.23 mmol with an L/P ratio of 12.3 (fig. 1A). This increase of maternal lactate and pyruvate is approximately equally divided between the period of labor and delivery (table III). The increase during delivery observed in the total population also occurred in those patients delivered by elective section (fig. 1A).

There is a wide range in respect to the increase of lactate and pyruvate in the individual patient. The patients were divided into three groups based upon the maternal lactate quartiles at delivery to demonstrate the range of individual behavior (fig. 1B). Group 1 includes 228 patients with a lactate concentration at delivery less than 1.46 mmol. Group 2 includes 457 patients with a lactate concentration between 1.46 and 2.60 mmol. Group 3 includes 226 patients with a lactate concentration at delivery greater than 2.60 mmol. Group 1 exhibits no increase of lactate and pyruvate during labor and delivery. Group 2 exhibits an average increase of lactate 1.1 mmol and pyruvate 0.10 mmol with an L/P ratio of 12.1 at delivery. Group 3 exhibits an average increase of lactate 3.0 mmol and pyruvate 0.19 mmol with an L/P ratio of 14.5 at delivery.

Measures of fetal lactate and pyruvate at the onset of labor are not available in this study or the literature. The estimated measures used in the study, lactate 1.7 mmol and pyruvate 0.14 mmol were derived from observations in the fetal umbilical artery in the elective section group. Based upon this assumption, the increase during labor and delivery was 2.03 mmol for lactate and 0.07 mmol for pyruvate based on fetal umbilical artery figures of 3.73 mmol for lactate and 0.21 mmol for pyruvate with an L/P ratio of 18.3 at delivery (fig. 2A).

The individual fetus as in the case of the mother exhibits a wide range in respect to the apparent increase of lactate and pyruvate during labor and delivery. The patients were again divided into three groups based upon the fetal umbilical artery lactate quartiles to demonstrate this range of individual behavior (fig. 2B). Group 1 includes 167 patients with a fetal umbilical artery lactate less than 2.17 mmol. Group 2 includes 334 patients with a fetal umbilical artery lactate between 2.17 and 3.62 mmol. Group 3 includes 172 patients with a fetal umbilical artery lactate greater than 3.62 mmol. Group 1 exhibits a slight increase of lactate and pyruvate during labor and delivery. Group 2 exhibits an average increase of lactate 1.7 mmol and pyruvate 0.06 mmol with L/P ratio 17.6 at delivery. Group 3 exhibits an average increase of lactate 4.6 mmol and pyruvate 0.13 mmol with an L/P ratio of 24.6 at delivery.

Mechanisms Accounting for Increased Maternal and Fetal Lactate

The average increase of lactate in the mother, 1.4 mmol and fetus, 2.0 mmol is due either to an increase of pyruvate or tissue hypoxia. The relative increase of lactate and pyruvate in the mother and fetus, assuming a constant lactate-pyruvate ratio of 10 with adequate tissue oxygenation, is presented in figure 3. These relationships suggest that in the mother approximately 64% of the increase of lactate is associated with an increase of pyruvate and 36% is independent of pyruvate, while in the

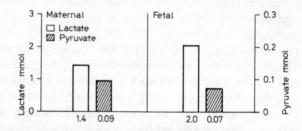

Fig. 3. The relative increase of maternal and fetal lactate and pyruvate during labor and delivery in the total population.

Table IV. The relative role of increased pyruvate and tissue hypoxia accounting for the increase of maternal lactate concentration in the total population and maternal group 2 and group 3, and for the increase of fetal lactate concentration in the total population and fetal group 2 and group 3

Mechanisms of lactate increase	Increase of pyruvate, %	Tissue hypoxia, %
Maternal		
Total population	64	36
Group 2	90	10
Group 3	60	40
Fetal		
Total population	34	66
Group 2	33	67
Group 3	33	67

Table V. The correlation coefficients between glucose, carbon dioxide tension, abnormal labor with pyruvate concentration and between oxygen tension and pyruvate with lactate concentration in maternal venous blood and fetal umbilical artery blood derived from a multiple regression analysis

	Maternal pyruvate		Fetal pyruvate	
	correlation	p value	correlation	p value
Glucose	0.369	<0.001	0.347	<0.001
CO_2 tension	0.376	<0.02	0.369	<0.001
Labor	0.429	<0.001	0.512	<0.001
	Maternal lactate		Fetal lactate	
	correlation	p value	correlation	p value
O_2 tension	0.195	<0.001	0.179	<0.001
Pyruvate	0.815	<0.001	0.760	<0.001

fetus approximately 34% of the increase of lactate is associated with an increase of pyruvate and 66% is independent of pyruvate. The relative increase of lactate and pyruvate in maternal group 2 and 3 and fetal group 2 and 3 is summarized in table IV. The increase of lactate associated with an increase of pyruvate in the mother was greater in group 2 than group 3 but in the fetus was essentially the same in group 2 and 3.

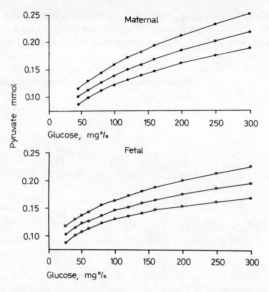

Fig. 4. Regression curves between glucose and pyruvate concentrations in maternal venous and fetal umbilical artery blood with 95% confidence limits.

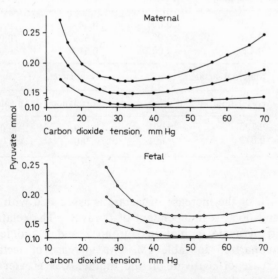

Fig. 5. Regression curves between carbon dioxide tension and pyruvate concentration in maternal venous and fetal umbilical artery blood with 95% confidence limits.

Multiple regression was used to assess those factors which may be contributing to the increase of maternal and fetal pyruvate. In the regressions, the logarithms of the variables were used because this was statistically more valid and gave better results. Regressions were performed using log pyruvate vs dependent variables in both maternal and fetal blood. The results are summarized in table V. Pyruvate was regressed against glucose, carbon dioxide tension and the type of labor. Glucose was a linear relation while carbon dioxide tension was a quadratic relation. The types of labor were two variables, i.e. normal and abnormal as deviations from no labor.

The correlation coefficients quoted in table V are coefficients of multiple correlation, i.e. the square root of the usual R-squared statistic. The coefficients opposite each variable relates to the model which includes that variable and the one above it. Thus, 0.376 is the multiple correlation coefficient from the model with the variables glucose and carbon dioxide tension. The p values are the significance probabilities for testing the inclusion of each set of variables in addition to the ones above it in table V. The computer program used was the SPSS program. Figures 4 and 5 summarize the associations found. In each case, the curve gives the dependent variable (pyruvate) vs one of the independent variables. The other variables are held at some typical value. The 95% confidence limits are given for each curve as an aid in judging the strengths of each relation.

There is a relatively strong association between glucose concentration and pyruvate concentration with a correlation coefficient in maternal blood of 0.37 and in fetal blood of 0.35. The regression curve, figure 4, suggests that any increase of maternal and fetal glucose concentration may be associated with an increase of pyruvate concentration. Pyruvate concentrations are greater than maternal and fetal baseline values when glucose concentrations exceed approximately 100 mg% which occurred in 52% of the maternal and 38% of the fetal observations.

There is a significant but very modest association between carbon dioxide tension and pyruvate concentration with an increase of the correlation coefficient in maternal blood from 0.37 to 0.38 and in fetal blood from 0.35 to 0.37. The regression curve, figure 5, indicates a relationship between decreasing maternal carbon dioxide tension and increasing pyruvate concentration when the maternal pCO_2 was less than 25 mm Hg which occurred in 9% of the maternal observations. Similarly, there is a relationship between decreasing fetal carbon dioxide tension and increasing pyruvate concentration when the fetal pCO_2 was less than 38 mm Hg which occurred in 11% of the fetal observations.

The relationship between each maternal medical, obstetrical and fetal

Table VI. The relationship between labor and increasing maternal venous and fetal umbilical artery pyruvate concentrations (p <0.001)

Labor	Quart 1 N-228	Quart 2 N-226	Quart 3 N-233	Quart 4 N-226
Maternal pyruvate				
None	41	27	19	3
Normal	141	145	153	136
Abnormal	46	54	61	87
	Quart 1 N-167	Quart 2 N-169	Quart 3 N-165	Quart 4 N-172
Fetal pyruvate				
None	56	11	7	6
Normal	80	116	113	108
Abnormal	31	42	45	58

complication with pyruvate concentration in maternal and fetal blood was analyzed. No significant relationships were identified with the exception of labor. The cross tabulations in table VI demonstrate that the increase of pyruvate in both maternal and fetal blood is associated with an increased incidence of abnormal labor. The significant association between abnormal labor and pyruvate concentration was demonstrated in the multiple regression analysis with an increase of the correlation coefficient in maternal blood from 0.38 to 0.43 and in fetal blood from 0.37 to 0.51.

Multiple regression analysis was also used to assess the association of oxygen tension as well as pyruvate with lactate concentration (table V). There is a significant association between oxygen tension and lactate with a correlation coefficient in maternal blood of 0.19 and in fetal blood of 0.18. The regression curve, figure 6, demonstrates a relationship between decreasing maternal oxygen tension with a small increase of lactate concentration. Similarly, there is a relationship between decreasing fetal oxygen tension and increasing lactate concentration with, in this instance, a more striking increase of lactate with the decrease of fetal oxygen tension. There is also a very strong association between pyruvate and lactate concentration with an increase of the correlation coefficient in maternal blood from 0.19 to 0.81 and in fetal blood from 0.18 to 0.76.

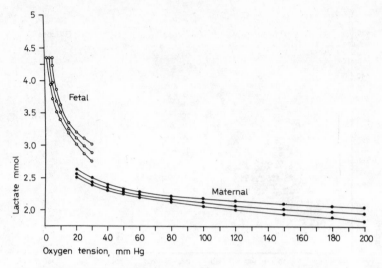

Fig. 6. Regression curves between oxygen tension and lactate concentrations in maternal venous and fetal umbilical artery blood with 95% confidence limits.

Maternal-Fetal Lactate and Pyruvate Relationships

There is a relationship between maternal and fetal concentrations of glucose, pyruvate and lactate. The correlation coefficients between maternal venous blood and fetal umbilical artery blood concentrations were for glucose 0.80, pyruvate 0.53, and lactate 0.36. These relationships for glucose and lactate are demonstrated in the scattergrams in figure 7.

The maternal-fetal gradients for glucose, pyruvate, and lactate for the three maternal lactate groups are presented in figure 8. Whereas the maternal-fetal glucose gradient remains constant in the three groups, the fetal-maternal gradient for pyruvate in group 1 is reversed in group 3, and fetal-maternal gradient for lactate decreases between group 1 and group 3.

The maternal and fetal lactate and pyruvate observations during labor and at delivery in 'normal' pregnancies are compared to 'complicated' pregnancies in figure 9. Although the increase of lactate in maternal venous blood during labor and delivery is essentially the same in the two groups, in the fetus, the increase of lactate in the 'complicated' pregnancies is significantly greater than the 'normal' pregnancies (p < 0.001).

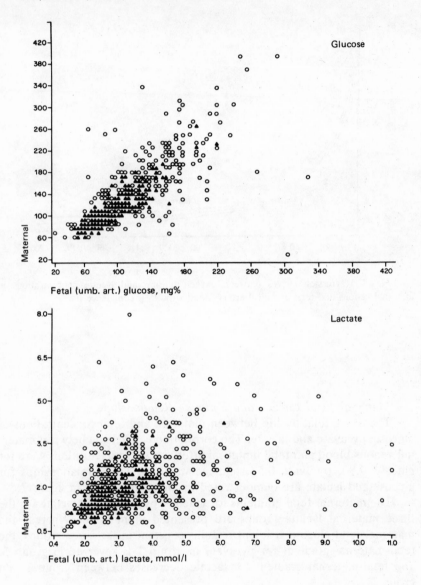

Fig. 7. Relationship between maternal venous and fetal umbilical artery glucose concentration and lactate concentration in the total population.

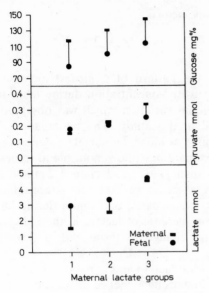

Fig. 8. Maternal venous-fetal umbilical artery gradients at delivery for glucose, pyruvate, and lactate concentrations.

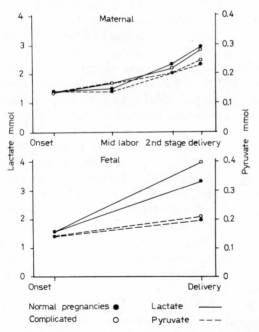

Fig. 9. Maternal and fetal lactate and pyruvate values during labor and at delivery in the 'normal' pregnancies and the 'complicated' pregnancies.

Discussion

The present review confirms the pattern of a *modest* increase of maternal and fetal lactate and pyruvate concentration during labor and delivery but emphasizes the individual variation which was observed in the spectrum of patients included in this study. The maternal venous blood measures obtained at the onset of labor, i.e. lactate 1.4 mmol and pyruvate 0.14 mmol, closely approximate baseline venous blood values in normal adults reported by HUCKABEE [19], i.e. lactate 1.2 mmol and pyruvate 0.12 mmol. The average increase of maternal venous blood lactate, 1.4 mmol, occurred in two equal parts, 0.75 mmol during labor and 0.65 mmol during delivery. The increase of lactate in the individual patient ranges from those with no change to those with a threefold increase of lactate concentration.

Although precise measures of fetal lactate and pyruvate at the onset of labor are not available, there is increasing evidence to suggest that the estimated values used in this study, i.e. lactate 1.7 mmol and pyruvate 0.14 mmol, are close approximations. There is increasing uniformity in resting umbilical artery lactate values in chronically cannulated fetal lamb, as reported by BURD et al. [7], 2.05 mmol, and ROBILLARD et al. [31], approximately 1.5 mmol, and in fetal monkey as reported by CHAR and CREASEY [9], 1.99 mmol. Human fetal capillary blood lactate measured in normal pregnancies later in the first stage of labor reported by SCHMID [33] were 1.97 mmol. Based upon these estimated values, there has been an average increase of fetal lactate of approximately 2.0 mmol with again a wide variation in the individual fetus from those exhibiting little change to those with a fourfold increase of lactate concentration.

The increase of lactate concentration in the mother and the fetus is due to an increase of pyruvate and tissue hypoxia acting simultaneously. The role of an increase of pyruvate appears to be relatively greater in the mother accounting for approximately two thirds of the increase of lactate in maternal venous blood, whereas in the fetus, it accounts for approximately one third of the increase of lactate during labor and delivery. The role of tissue hypoxia in the mother is principally apparent when the increase of lactate is marked, i.e. maternal group 3, whereas in the fetus it appears to have been uniformly responsible for two thirds of the increase of lactate during labor and delivery.

The observations in this study indicate not only that an increase of pyruvate is a mechanism accounting for the increase of maternal and fetal lactate during labor and delivery, but that those factors which have been demonstrated to account for a parallel increase of pyruvate and lactate in animals and man, hyperglycemia, hypocapnea, and catecholamines are also relevant to the increase of pyruvate in the obstetric patient and fetus. Hyperglycemia leads to increased glycolysis with an increase of pyruvate with a parallel increase of lactate concentration. This response to hyperglycemia has been demonstrated in the adult [6, 19], the fetal lamb [34, 31], and the human fetus [13, 32]. Hyperventilation with hypocapnea and a respiratory alkalosis leads to a compensatory increase of pyruvate and lactate which may be marked [19] and, if sustained, the increase will continue until a metabolic acidosis develops [11]. Catecholamines have long been known to increase lactate concentration [8, 14, 15] presumably due to an increase of pyruvate concentration [19].

Maternal and fetal hyperglycemia has frequently accounted for a moderate increase of pyruvate concentration during labor and delivery. Maternal and fetal glucose concentrations in excess of 100 mg%, which are associated with an increase of pyruvate concentration, occurred in approximately 50% of maternal and 40% of fetal observations. Maternal and fetal hypocapnea has played a small but significant role in respect to the increase of pyruvate concentration during labor and delivery. Maternal pCO_2 less than 25 mm Hg and fetal pCO_2 less than 38 mm Hg, which is associated with a modest increase of pyruvate concentration, occurred in approximately 10% of maternal and fetal observations. There is a significant relationship between abnormal labor and increased maternal and fetal pyruvate concentration. There is evidence that abnormal labor is associated with increased production of catecholamines in the mother and fetus [4, 39]. It is proposed that increased maternal and fetal catecholamines may be the factor in abnormal labor accounting for the increase of pyruvate concentration.

The second mechanism accounting for the increase of maternal and fetal lactate is tissue hypoxia in which the rate of oxygen supply to the interior of the cell has fallen below the rate of energy utilization, i.e. below the oxygen requirement. This is a relative oxygen lack and the oxidized DPN produced by the lactic dehydrogenase system replaces in part the oxidation of the DPN by molecular oxygen. The factors leading to tissue hypoxia are complex. General or regional circulatory failure with ischemia is likely the most potent mechanism leading to tissue hypoxia [20]. Hypoxemia is a cause of tissue hypoxia but may be modified by a compensatory increased blood flow. Hypoxemia, although only one part of the mechanism of tissue hypoxia, both in maternal and fetal blood, demonstrates a significant relationship to increasing lactate concentration in this study. The decrease of maternal oxygen tension, over the wide range observed, is associated with a small increase of lactate concentration whereas the decrease of fetal oxygen tension, which is over a smaller but in relation to the oxygen dissociation curve a more critical range, is associated with a more marked increase of lactate concentration.

Animal studies indicate that fetal lactate does not usually cross the placenta from the fetus to the mother. Huckabee et al. [22], in 1962, proposed that the placenta could take up lactate from the fetal circulation and placental oxygen serve to supplement fetal oxygen requirements. Otey et al. [28], in 1964, in view of the umbilical artery-vein and lactate gradient noted in most human studies, suggested the placenta served as a buffer for the fetus. The recent evidence that the placenta provides the fetus with lactate, which is the source of 25% of the energy requirements of the fetus [7, 9] indicates that the placenta has a further role in the regulation of fetal lactate metabolism.

The significant correlation between maternal and fetal concentrations of glucose, pyruvate, and lactate observed in this study may be due to placental transfer or parallel metabolic changes in the mother and the fetus. The high degree of correlation between maternal and fetal glucose concentrations is a result of the efficient transfer of glucose from mother to fetus which has been demonstrated in animals [3, 17, 34] with similar maternal-fetal gradients reported in humans [1, 13, 27, 30, 35].

The relationships in respect to pyruvate and lactate are quite different from those for glucose. The maternal-fetal correlations for pyruvate and lactate are not the same order as glucose and there is no uniformity of maternal-fetal gradients with increasing maternal concentrations. The relative role of the mechanisms accounting for increased lactate, i.e. increased pyruvate and tissue hypoxia, in the mother and fetus are different. When the 'normal' and 'complicated' pregnancies are compared, whereas the increase of lactate in the mother is similar in the two groups, there is a significantly greater increase of lactate in the fetus of the 'complicated' pregnancies. These observations do not rule out placental transfer; however, they are in keeping with concept that the increase of maternal and fetal pyruvate and lactate is due to independent increases in the mother and fetus and, as proposed from animal studies, little, if any, pyruvate and lactate has crossed the placenta particularly from the fetus to the mother.

What is the clinical significance of these observations? Although metabolic acidosis remains the best indicator of fetal hypoxia with tissue oxygen debt, there are several points relevant to the interpretation of these measures. Some of the principle fixed acid-lactate accounting for this metabolic acidosis is due to an increase of pyruvate in addition to that due to tissue oxygen debt. The significance of this component can be established by measures of pyruvate and lactate and estimation of the L/P ratio. The increase of lactate concentration due to pyruvate may in the future be avoided in part as the understanding of the mechanisms leading to an increase of maternal and fetal pyruvate develops. Finally, the resolution of an episode of fetal metabolic acidosis *in utero* will be slow in view of the increasing evidence of the limited placental transfer of fetal lactate to the maternal circulation and the delayed excretion of such fixed acids by the fetal kidney.

Summary

The average increase during labor and delivery of maternal lactate was 1.4 mmol and fetal lactate 2.0 mmol with a wide range of increase in the individual patient and fetus. The increase of lactate in the mother and fetus is due to an increase of pyruvate and tissue hypoxia acting simultaneously. The factors accounting for the increase of pyruvate in the mother and fetus include hyperglycemia, hypocapnea, and abnormal labor possibly due to increased catecholamines. The observations in this study are in keeping with the concept that fetal lactate does not cross the placenta.

References

1 ANDERSON, G.; CORDERO, L., and HON, E. H.: Hypertonic glucose infusion during labor. Obstet. Gynec., N.Y. *36:* 405 (1970).
2 BARKER, S. B. and SUMMERSON, W. H.: The colorimetric determination of lactic acid in biological material. J. biol. Chem. *138:* 535 (1941).
3 BATTAGLIA, F. C.; HELLEGER, A. E.; HELLER, C. R. J., and BEHRMANN, R.: Glucose concentration gradients across the maternal surface, the placenta, and the ammion of the rhexes monkey. Am. J. Obstet. Gynec. *88:* 32 (1964).
4 BRENNER, W. E.; OGBURN, P. L.; DINGFELDER, J. R.; STANROWSKY, L. G., and ZUSPAN, F. P.: Catecholamines during therapeutic abortion induced with intra-amniotic prostaglandin F_2. Am. J. Obstet. Gynec. *130:* 178 (1978).
5 BRITTON, H. G.; NIXON, D. A., and WRIGHT, G. H.: The effect of acute hypoxia on the sheep fetus and some observations on recovery from hypoxia. Biol. Neonate *11:* 277 (1967).
6 BUEDING, E. and GOLDFARB, W. J.: Blood changes following glucose, lactate and pyruvate injections in man. J. biol. Chem. *147:* 33 (1943).
7 BURD, L. I.; JONES, M. D.; SIMMONS, M. A.; MAKOWSKI, E. L.; MESCHIA, G., and BATTAGLIA, F. C.: Placental production and foetal utilization of lactate and pyruvate. Nature, Lond. *254:* 710 (1975).
8 CORI, C. F. and CORI, G. T.: The mechanism of epinephrine action. IV. The influence of epineplurine on lactic acid production and blood sugar utilization. J. biol. Chem. *84:* 683 (1929).
9 CHAR, V. C. and CREASEY, R. K.: Lactate and pyruvate as fetal metabolic substrates. Pediat. Res. *10:* 231 (1976).
10 DEROM, R.: Anaerobic metabolism in the human fetus. I. The normal delivery. Am. J. Obstet. Gynec. *89:* 241 (1964).
11 EICHENHOLZ, A.; MULHAUSEN, R. O.; ANDERSON, W. E., and MACDONALD, F. M.: Primary hypocapnoea a cause of matabolic acidosis. J. appl. Physiol. *17:* 283 (1962).

12 FRIEDMAN, E. A.; GRAY, M. J.; GRYNFOGEL, M.; HUTCHINSON, D. L.; KELLY, W. T. and PLEUTH, A. A.: The distribution and metabolism of C^{14} labelled lactic acid and bicarbonate in pregnant primates. J. clin. Invest. *39:* 227 (1960).

13 GARDMARK, S.; GENNSER, G.; JACOBSON, G.; ROOTH, G., and THORELL, J.: Influence on fetal carbohydrate and fat metabolism and on acid-base balance of glucose administration to the mother during labor. Biol. Neonate *26:* 129 (1975).

14 GINSBURG, J. and JEACOCK, M. K.: Effect of adrenaline on human placental lactate production *in vitro.* Am. J. Obstet. Gynec. *90:* 794 (1964).

15 GRIFFITH, E. R.; OMACHI, A.; LOCKWOOD, J. E., and LOOMIS, T. A.: The effect of intravenous adrenolin on blood flow, sugar retention, lactate output and respiratory metabolism of the peripheral tissues in the anaesthetized cat. Am. J. Physiol. *149:* 49 (1947).

16 HENDRICKS, C. H.: Studies on lactic acid metabolism in pregnancy and labor. Am. J. Obstet. Gynec. *73:* 492 (1957).

17 HINCHLEY, C. M.; STENGER, V. G., and BLECHNER, J. N.: Glucose concentration gradients across the uterus and placenta of the pregnant primate. Am. J. Obstet. Gynec. *104:* 893 (1969).

18 HUCKABEE, W. E.: Control of concentration gradients of pyruvate and lactate across all membranes in blood. J. appl. Physiol. *9:* 163 (1956).

19 HUCKABEE, W. E.: Relationships of pyruvate and lactate during anaerobic metabolism. 1. Effects of infusion of pyruvate or glucose and of hyperventilation. J. clin. Invest. *37:* 244 (1958).

20 HUCKABEE, W. E.: Relationships of pyruvate and lactate during anaerobic metabolism. III. Effect of breathing low oxygen gases. J. clin. Invest. *37:* 264 (1958).

21 HUCKABEE, W. E.; METCALFE, J.; PRYSTOWSKY, H., and BARRON, D. H.: Movement of lactate and pyruvate in pregnant uterus. Am. J. Physiol. *202:* 193 (1962).

22 HUCKABEE, W. E.; METCALFE, J.; PRYSTOWSKY, H., and BARRON, D. H.: Insufficiency of O_2 supply to pregnant uterus. Am. J. Physiol. *202:* 198 (1962).

23 LOW, J. A.; PANCHAM, S. R.; WORTHINGTON, D., and BOSTON, R. W.: Acid-base, lactate and pyruvate characteristics of the normal obstetric patient and foetus during the intrapartum period. Am. J. Obstet. Gynec. *120:* 862 (1974).

24 LOW, J. A.; PANCHAM, S. R.; WORTHINGTON, D., and BOSTON, R. W.: The acid-base and biochemical characteristics of intrapartum fetal asphyxia. Am. J. Obstet. Gynec. *121:* 446 (1975).

25 MANN, L. I.: Effects in sheep of hypoxia on levels of lactate, pyruvate and glucose in blood of mother and fetus. Pediat. Res. *4:* 46 (1970).

26 MARX, G. F. and GREENE, N. H.: Maternal lactate, pyruvate and excess lactate production during labor and delivery. Am. J. Obstet. Gynec. *90:* 786 (1964).

27 OAKLEY, N. W.; BEARD, R. W., and TURNER, R. C.: Effect of sustained maternal hyperglycemia in the fetus in normal and diabetic pregnancies. Bri. med. J. *i:* 466 (1972).

28 OTEY, E.; STENGER, V.; EITZMAN, D.; ANDERSON, T.; GESSNER, I., and PRYSTOWSKY, H.: Movement of lactate and pyruvate in the pregnant uterus of the human. Am. J. Obstet. Gynec. *90:* 747 (1964).

29 OTEY, E.; STENGER, V.; EITZMAN, D., and PRYSTOWSKY, H.: Further observations on the relationship of pyruvate and lactate in human pregnancy. Am. J. Obstet. Gynec. *97:* 1076 (1967).

30 PATERSON, P.; PHILLIPS, L., and WOOD, C.: Relationship between maternal and fetal blood glucose during labor. Am. J. Obstet. Gynec. *98:* 938 (1967).

31 ROBILLARD, J. E.; SESSIONS, C.; KENNEDY, R. L., and SMITH, F. G.: Metabolic effects

of constant hypertonic glucose infusion in well oxygenated foetuses. Am. J. Obstet. Gynec. *130:* 199 (1978).

32 SABATA, V.; ZNAMENACEK, K.; PRIBYLOV, H., and MELICHAR, V.: The effects of glucose infusion administered to the mother during delivery upon the metabolism of the premature newborn; in HORSKY and STEMBARA Intrauterine changes to the fetus, pp. 435–439 (Exerpta Medica, Amsterdam 1967).

33 SCHMID, J.: Glucose, lactate and pyruvate in pregnancy and during birth. Adv. Obstet. Gynaec., vol. 50, pp. 1–99 (Karger, Basel 1973).

34 SHELLEY, H. J.: The case of chronically catheterized lambs for the study of foetal metabolism; in COMLINE, CROSS, DAWES, and NATHANIELSZ Foetal and neonatal physiology, pp. 360–397 (Cambridge University Press, Cambridge 1973).

35 SPELLACY, W. N.; GROETZ, F. C.; GREENBURG, B. Z., and ELLIS, J.: The human placental gradient for plasma insulin and blood glucose. Am. J. Obstet. Gynec. *96:* 698 (1964).

36 STEELE, S. M.; JACKSON, G. B., and WOLKOFF, A. H.: Some aspects of blood lactate levels in mother and fetus. Am. J. Obstet. Gynec. *105:* 569 (1969).

37 STEMBERA, Z. K. and HODR, J.: The relationship between the levels of glucose, lactic acid and pyruvic acid in the mother and umbilical vessels of the healthy fetus. Biol. Neonate *10:* 227 (1966).

38 WARNES, D. M.; SEAMARK, P. F., and BALLARD, F. J.: Metabolism of glucose fructose and lactate *in vivo* in chronically cannulated foetuses and in suckling lamb. J. Biochem. *162:* 617 (1977).

39 ZUSPAN, F. P.: Adrenal gland and sympathetic nervous system response in eclampsia. Am. J. Obstet. Gynec. *114:* 304 (1972).

Dr. J. A. LOW, Department of Obstetrics and Gynaecology of Queen's University, *Kingston, Ontario K7L 3N6* (Canada)

Lactate in Acute Conditions. Int. Symp., Basel 1978, pp. 48–55 (Karger, Basel 1979)

Cord-Blood Studies

P.-J. Ditesheim and H. Bossart

Centre Hospitalier Universitaire Vaudois, Service d'obstétrique et gynécologie, Lausanne

Introduction

There exists a wide gap between a theoretical knowledge of lactate metabolism and practical knowledge in the human fetus during delivery and its first minutes of life. We know that lactate passes from mother to placenta, fetus and back, but we do not know what happens in different organs, what perfusion and utilization rates are, etc. [4].

Every clinical situation – condition of the mother, gestational age, maturity of the fetus, its well-being – all these factors and many more change and therefore produce a tremendous amount of individual metabolic situations.

We have therefore chosen one single and simple question to be discussed at this symposium: Are there significant relationships between lactate gradients in cord blood and other acid-base parameters? Only if such correlations can be found can we continue to look for possible clinical significance.

Material and Methods

Maternal venous blood is collected at the moment of delivery from a brachial vein without any compression. The cord is clamped immediately after delivery and blood is collected from vein and artery in heparinized syringes. Lactate is determined either immediately in total blood or later in plasma which is obtained by immediate centrifugation. Lactate is determined with the Roche electro-chemical equipment [1, 3]. Routine acid-base parameters are obtained immediately after sampling by means of a Radiometer acid-base laboratory ABL 1.

After elimination of all technically non-perfect cases, there remained 172 observations for statistical evaluation. Two categories were created (table I): first category – 100 cases with less than 4 mmol/l of lactate in the umbilical artery, and second category – 72 cases with 4 and more mmol/l of arterial lactate.

Table I. Breakdown of cases distributed in low and high-lactate groups. L_A = Lactate in umbilical artery

		$L_A < 4.0$ mmol/l: 1st category, n	$L_A \geqslant 4.0$ mmol/l: 2nd category, n
Cumulative Apgar score $\geqslant 27 (\geqslant 8/\geqslant 9/10)$	Spontaneous vertex deliveries	90	57
	Spontaneous breech delivery	1	—
	Forceps deliveries	3	1
		94 (94%)	58 (81%)
Cumulative Apgar score <27	Spontaneous vertex deliveries	4	8
	Forceps deliveries	1	2
	Caesarean section for intrauterine growth retardation	1	2
	Spontaneous breech deliveries	—	2
		6 (6%)	14 (19%)
		100	72

Roughly speaking, we have more clinical problems (cumulative Apgar score less than 27) in the high lactate category. The remaining high lactate cases (about 80%) probably correspond more or less to the newborn having received lactate from the mother and the placenta without excessive fetal production [2].

Results

Lactate in umbilical vein and artery is correlated successively with umbilical and maternal pH, base deficit, pO_2 and pCO_2.

The 100 normal cases are again divided in two groups, one with arterio-venous lactate gradient, the other with venous-arterial gradient. This group selection shall be respected in all our observations for the normal and the high-lactate category.

The mean values in both groups of cord blood do not differ significantly; but there is a significant difference in maternal venous blood. This fact is probably in relation to lactate transfer from mother to fetus (table II).

In this high-lactate category (table III), there is a significant difference between the two groups in maternal venous and umbilical arterial blood. In 31 cases (43%), the lactate gradient is positive for the umbilical vein,

Table II. Lactate mean values (mmol/l). Low-lactate category

	Means and standard deviations		t value
	group I lactate A > V	group II lactate A < V	
Lactate M	3.98 ± 1.43	5.51 ± 1.94	4.49*
Lactate A	3.89 ± 1.78	3.42 ± 1.02	−1.61
Lactate V	3.53 ± 1.41	3.68 ± 1.05	0.63
n	52	48	

Left column: arterio-venous gradient (LA > V); right column: venous-arterial gradient (LA < V). A = Blood from umbilical artery; V = blood from umbilical vein; M = maternal venous blood.
* $p < 0.01$.

Table III. Lactate mean values (mmol/l). High-lactate category

	Means and standard deviations		t value
	group I lactate A > V	group II lactate A < V	
LM	5.14 ± 1.82	6.77 ± 2.28	−3.18*
LA	6.38 ± 2.70	4.88 ± 0.88	2.96*
LV	5.59 ± 2.28	5.20 ± 0.94	0.88
n	41	31	

For abbreviations, see table II.
* $p < 0.01$.

Table IV. pH mean values in the low-lactate category

	Means and standard deviations		t value
	group I lactate A > V	group II lactate A < V	
pH M	7.318 ± 0.060	7.279 ± 0.071	−2.98*
pH A	7.225 ± 0.064	7.244 ± 0.041	1.76
pH V	7.302 ± 0.055	7.298 ± 0.046	0.37
n	52	48	

For abbreviations, see table II.
* $p < 0.01$.

Table V. pH mean values in the high-lactate category

	Means and standard deviations		t value
	group I lactate A > V	group II lactate A < V	
pH M	7.319 ± 0.069	7.286 ± 0.081	1.70
pH A	7.159 ± 0.116	7.224 ± 0.070	-2.71*
pH V	7.254 ± 0.106	7.289 ± 0.061	-1.63
n	41	31	

For abbreviations, see table II.
* p < 0.01.

and positive for the umbilical artery in 41 cases. pH mean values according to arterio-venous gradient and venous-arterial gradient are presented in the low-lactate category (table IV) and high-lactate category (table V).

When lactate is low, there exists a significant correlation in maternal venous blood between the two gradient groups. In the high-lactate group, we find significance in the umbilical artery group. Again, we cannot clearly say what this means, but we might consider the following reasons: In the low-lactate category, the most important lactate factor is the maternal metabolism, or in other, more clinical words, maternal fatigue, length of parturition, nutritional state and so forth. In presence of high-lactate levels are two important causes: the mother, as in the first category, and the fetus with its own endogenous lactate production. This explains significance in the umbilical arteries.

We present our material in the same way for base deficit (tables VI, VII). The same significant differences can be found as in the previous groups: in the low-lactate category, maternal venous blood difference; high-lactate category, umbilical artery difference. The same observations can be made concerning pCO_2 (tables VIII, IX) and pO_2 (tables X, XI).

In these last comparisons, we find a limit-significant difference in the umbilical arterial blood in the low-lactate category, but no significance in the high-lactate category. But we all know that pO_2 measurements are technically difficult and have quite a low precision and accuracy. Then of course we realize that maternal venous blood is basically inadequate, but for ethical reasons, we do not start arterial sampling unless the clinical condition of the mother gives us reason to do so. In obstetrics, such conditions are rare.

We nevertheless note a certain trend in the high-lactate category. pO_2

Table VI. Base deficit in the low-lactate category

	Means and standard deviations		t value
	group I lactate A > V	group II lactate A < V	
BD M	8.65 ± 2.38	9.83 ± 2.28	2.53*
BD A	7.76 ± 2.78	7.05 ± 1.57	−1.56
BD V	6.89 ± 2.18	6.99 ± 1.82	0.24
n	52	48	

For abbreviations, see table II.
* p > 0.05.

Table VII. Base deficit in the high-lactate category

	Means and standard deviations		t value
	group I lactate A > V	group II lactate A < V	
BD M	9.62 ± 2.36	10.67 ± 2.97	−1.52
BD A	10.35 ± 4.22	8.32 ± 2.00	2.48*
BD V	9.06 ± 3.46	8.08 ± 1.76	1.43
n	41	31	

For abbreviations, see table II.
* p < 0.05.

Table VIII. pCO_2 in the low-lactate category

	Means and standard deviations		t value
	group I lactate A > V	group II lactate A < V	
pCO_2M	32.65 ± 0.56	35.30 ± 0.75	2.02*
pCO_2A	50.12 ± 0.92	48.65 ± 0.65	−0.91
pCO_2V	39.27 ± 0.60	39.77 ± 0.66	0.40
n	52	48	

For abbreviations, see table II.
* p < 0.05.

Table IX. pCO$_2$ in the high-lactate category

	Means and standard deviations		t value
	group I lactate A > V	group II lactate A < V	
pCO$_2$M	31.01 ± 5.95	31.71 ± 6.54	−0.43
pCO$_2$A	55.27 ± 16.26	47.14 ± 10.69	2.42*
pCO$_2$V	41.94 ± 10.51	38.40 ± 7.16	1.58
n	41	31	

For abbreviations, see table II.
* p < 0.05.

Table X. pO$_2$ in the low lactate category

	Means and standard deviations		t value
	group I lactate A > V	group II lactate A < V	
pO$_2$M	51.57 ± 20.13	52.24 ± 17.53	0.18
pO$_2$A	19.22 ± 4.80	21.13 ± 3.37	2.29*
pO$_2$V	29.21 ± 5.47	30.03 ± 4.41	0.83
n	52	48	

For abbreviations, see table II.
* p < 0.05.

Table XI. pO$_2$ in the high lactate category

	Means and standard deviations		t value
	group I lactate A > V	group II lactate A < V	
pO$_2$M	52.93 ± 20.97	53.28 ± 24.04	−0.06
pO$_2$A	16.27 ± 5.07	18.29 ± 4.55	−1.70
pO$_2$V	27.19 ± 6.20	28.47 ± 5.40	−0.89
n	41	31	

For abbreviations, see table II.

Table XII. High-lactate category: Correlations between lactate in umbilical artery, pH, base deficit and pO_2 in umbilical artery. Significance is only found for base-deficit correlations

	Lactate umbilical artery			Lactate umbilical vein	
	group I lactate A>V	group II lactate A<V		group I lactate A>V	group II lactate A<V
pH A	−0.20	−0.04	pH V	−0.07	−0.52*
BD A	−0.74*	0.44*	BD V	0.86*	0.54*
pO$_2$A	−0.19	−0.06	pO$_2$V	−0.19	−0.02

* $p < 0.01$.

is higher in all compartments in the venous-arterial gradient group, indicating that fetal oxygen lack is more likely to exist with a positive gradient from umbilical artery to vein.

After having discussed lactate *gradients* only, we will now draw a few correlations between lactate *levels* and other acid-base parameters. If one correlates lactate in the umbilical artery to pH values in the different compartments, there is basically good correlation only in the group with positive arterio-venous gradient ($p < 0.001$). pO_2 is significantly correlated only to the group with arteriovenous gradients ($p < 0.034$). Positive correlation between lactate and pCO_2 exists only if lactate is higher in the umbilical artery ($p < 0.01$).

Finally, there is good correlation in all compartments and situations between lactate and base deficit; a result which can be easily understood ($p < 0.001$).

We finally come to some correlations in the high-lactate category, the correlation between lactate in umbilical artery and vein, subdivided in the two gradient groups and correlated to pH, base deficit and pO_2. Results are expressed in correlation coefficient r.

Correlations are significantly related to base deficit, the same finding as in the low-lactate category (table XII). The only other significant relation is found between lactate in umbilical vein with positive venous gradient and umbilical vein pH.

It is of course rather depressing to state that high lactate levels correlate poorly with other important acid-base parameters. A disturbing factor in this category of high lactate is of course the fact that roughly only 20% of the cases showed a certain amount of fetal distress. Because of this lack of fetal pathology, we refrained from creating a special homogenous group with fetal distress cases only. If we could do so, we could certainly expect better correlations.

Discussion

Let us stress a few considerations that we made before starting our cord-blood determinations. (1) In clinical obstetrics, we badly need biochemical parameters directly involved in the mechanism of fetal hypoxia. (2) These parameters must be obtained at the bedside; in other words, obtained in a few minutes. (3) The technique must be practicable within the labor ward and by what we may call 'normal' staff members. These conditions seemed to be realized with lactate and its new electrochemical determination.

Unfortunately, cord blood shows a metabolic pattern of very complicated origin. Mother, placenta, and fetus have their own metabolism and influence themselves mutually. Long-lasting factors depending on fetal growth, maternal health, nutritional habits, etc. mixed with acute factors, depending on uterine basal tone and contractions, maternal blood pressure and fetal response to all this.

Cord blood is, so to say, one picture out of a film. But still, cord blood reflects to a certain extent the metabolic situation at the beginning of extra-uterine life.

The relatively poor correlation that we found indicates that further research has to be made by a dynamic approach. Lactate build-up and removal related to time and clinical situations will be the clue to better understanding.

References

1 DITESHEIM, P.-J. et BOSSART, H.: Premiers essais de mesures du L-lactate plasmatique au moyen du 'Lactate-Analyzer 5400'. Schweiz. med. Wschr. *106:* 1598–1601 (1976).
2 LOW, J. A.; PANCHAM, S. R.; WORTHINGTON, D., and BOSTON, R. W.: Acid-base, lactate and pyruvate characteristics of the normal obstetric patient and fetus during the intrapartum period. Am. J. Obstet. Gynec. *120:* 862–867 (1974).
3 SCHELLENBERG, J.-C.; JANECEK, P.; DITESHEIM, P.-J., and BOSSART, H.: Instantaneous lactate determination with the 'Lactate-Analyzer 5400' in cord blood and maternal venous blood and its correlation to other acid-base parameters and clinical status of the new-born. Free communication. 5th European Congr. Perinatal Medicine, Uppsala 1976.
4 SCHMID, J.: Glukose, Laktat und Pyruvat in der Schwangerschaft und unter der Geburt. Adv. Obstet. Gynaec., vol. 50, p. 1 (Karger, Basel 1973).

Dr. P.-J. DITESHEIM, Service d'obstétrique et gynécologie, Hôpital de district, *CH-1260 Nyon* (Switzerland)

Lactate in Acute Conditions. Int. Symp., Basel 1978, pp. 56–68 (Karger, Basel 1979)

Arterial Lactate and Pyruvate Concentration in the Normal and Asphyxiated Newborn Infant[1]

LARS-ERIC BRATTEBY and STEN SWANSTRÖM

Perinatal Research Unit, Department of Pediatrics and Department of Clinical Physiology, University Hospital, Uppsala

Introduction

In their publication, *Acidose lactique et asphyxie du nouveau-né*, MATTHIEU *et al.* [8] described the correlation between perinatal asphyxia and blood lactate concentration in the newborn. They also demonstrated the prognostic value of lactate analysis in infants with hyaline membrane disease, and showed a correlation between lactate and several variables including base deficit, pH, arterial oxygen saturation, and degree of intrapulmonary shunting. From these studies, but also from studies by other investigators [3, 9, 10], it is obvious that the postnatal lactate concentration is correlated to different signs of perinatal hypoxia.

The results concerning the pyruvate concentration and the lactate/pyruvate ratio in normal and asphyxiated newborn infants are more conflicting. Higher pyruvate values and lactate/pyruvate ratios have been observed in severely depressed newborn infants than in normal neonates [3, 9], but no consistent relationship has been reported between postnatal pyruvate concentration or lactate/pyruvate ratio and degree of intrauterine asphyxia.

Even if a group of asphyxiated newborn infants could be discriminated from a group of normal infants by the lactate or pyruvate concentration or the lactate/pyruvate ratio, an unsolved question remains: How valid is the estimation of prenatal fetal hypoxia from these biochemical variables in the individual newborn? This is an important clinical question, since newborn infants may have been exposed to hypoxia without having a low Apgar score, while other newborns may be in a poor condition with a low Apgar score for other reasons than hypoxia. So there is obviously a need for an objective and quantitative measure of past hypoxia in the indi-

[1] Supported by grants from the Bank of Sweden Tercentenary Foundation.

vidual newborn. Does the lactate or pyruvate concentration, or the lactate/pyruvate ratio, fulfil the requirements of such a test? It is the aim of this paper to discuss this question on the basis of lactate and pyruvate concentrations during the first 2 h of life in two groups of infants, one reference group and one group of asphyxiated infants without additional postnatal hypoxia after resuscitation.

Material and Methods

The *reference group* of 59 newborn infants satisfied very strict criteria of normality (table I). Their mothers, of ages 18–35 years, were healthy and were in normal physical condition before and during pregnancy. 31 were primipara and 28 were multipara. They had normal vaginal deliveries which were electronically monitored. Immediately after birth, the infants were transferred to an incubator and were closely observed in thermoneutral

Table I. Clinical data of the reference group, 59 cases

A. *Labor and delivery*
1. Vaginal deliveries
2. Spontaneous start of labor
   Membranes ruptured <24 h before admission
   No hypertension or generalized edema on admission
3. Duration of first stage 8.8±3.3 h in primiparae
   Duration of first stage 5.8±2.3 h in multiparae
   Duration of second stage 58±36 min in primiparae
   Duration of second stage 19.6±19 min in multiparae
4. Normal fetal heart rate pattern (continuous electronic monitoring)
   Normal maternal blood pressure during delivery
   Normal fetal scalp pH>7.10
   (Mean fetal scalp pH 7.352 15 min before birth, n = 26)
   No meconium passage during labor

B. *Newborn infants*
1. Apgar score ≥8 at 1, 5, and 15 min of age
2. Gestational age 38–42 weeks
3. Birth weight (mean±1 SD) = 3.5±0.5 kg
4. Normal neonatal adaptation based on:
   Intermittent monitoring of arterial blood gases and arterial blood pressure
   Continuous monitoring of ECG
   Continuous monitoring of ventilation
   Continuous monitoring of movements
   Continuous clinical observation of alertness, tonus, movements, respiration, and skin color during the first 2 h of life
5. Normal first week of life
   Repeated clinical and neurological examinations
6. Normal development at 18 months of age

---

environment for 2 h. The criteria of normality in the newborn infants were based on the
Apgar score, arterial blood pressure, continuous monitoring of ECG, pulmonary ventilation,
motor activity and close clinical observations. During the first 2 h of life, these infants had
normal arterial blood gases and acid-base values [5, 10]. During the first week of life, they
were examined clinically and neurologically according to a strict program. At 18 months of
age, their development was normal as judged from their clinical history, careful clinical
examination, and psychological tests by the Griffith method [4].

The *asphyxia group* of 13 newborn infants was selected on the basis of one or more
clinical signs of intrauterine distress. They all had 1-minute Apgar scores between 1 and 6,
and all but two also showed a pathological fetal heart rate pattern and/or meconium
passage. No other cause of fetal or neonatal depression than hypoxia was suspected in these
infants. After resuscitation, they all had an adequate gas exchange and a normal clinical
recovery. No infant was intubated. The asphyxia group is presented in more detail in table
II, which gives some of the signs of intrauterine distress and the Apgar scores at 1, 5, and
15 min after birth. The scores at these time points were summed to form a cumulative
Apgar score.

After resuscitation, the asphyxiated infants were observed and monitored in the
incubator in a thermoneutral environment during the first 2 h of life according to the same
program as the normal infants.

Table II. Clinical data for the asphyxia group, 13 cases

Case No.	Asphyxia group	Birth weight g	Gest. age weeks	Apgar score			Cumulative Apgar score (sum: 1+5+15)
				1 min	5 min	15 min	
1	Breech delivery	3,220	43	1	4	7	12
2	Fetal bradycardia, emergency cesarean section	3,650	41	3	5	6	14
3	Breech delivery	3,360	40	2	7	7	16
4	Fetal bradycardia, meconium passage	3,270	42	3	6	9	18
5	Fetal bradycardia, emergency cesarean section	3,700	41	3	7	9	19
6	Fetal bradycardia	3,430	41	3	8	9	20
7	Fetal bradycardia	3,600	40	5	7	8	20
8	Meconium passage	3,000	42	4	7	10	21
9	Intrauterine growth retardation, cesarean section	2,100	35	5	8	10	23
10	Hypotonic uterine function, vacuum extraction	3,470	40	6	8	9	23
11	Silent fetal heart rate pattern, double footling	3,350	40	6	9	9	24
12	Fetal bradycardia	3,320	40	6	9	10	25
13	Vertex delivery	3,760	39	6	9	10	25

The parents were informed about the aim and procedure of the investigation and gave their full consent to participation.

In all these infants, the blood gases, acid-base balance, and lactate and pyruvate concentrations were determined from repeated blood samples taken from a catheter (Argyle 5, French) in the umbilical artery. Immediately after sampling, blood for lactate and pyruvate analyses was handled by a special technician and was mixed with ice-cold metaphosphoric acid. These analyses were performed by the enzymatic method of Marbach and Weil [7]. The analytical error for lactate was ±2.3% and for pyruvate +2.5%.

Results and Discussion

The mean values and standard deviations of the lactate and pyruvate concentrations and the lactate/pyruvate ratios at different time intervals during the first 2 h of life for the normal infants are presented in table III. In figure 1, the mean lactate values of the normal infants of the present study at different time intervals are compared with data presented earlier in the literature. Ignoring the differences between results of different investigations and looking at the general trend of these values, it is seen that there is an early, sharp increase in lactate concentration to a peak value about 10 min postnatally, followed by a rapid fall during the next 20 min. Then comes a more gradual decline. The lactate level normally seen in the adult is not attained in the normal infant until 2–3 days after birth.

Table III. Mean (±SD) arterial lactate and pyruvate concentrations and the lactate pyruvate ratio in the reference group

Reference group	Sampling times, minutes after birth					
	10	20	30	60	90	120
Lactate, mmol/l						
Mean	4.72	4.32	3.70	2.78	2.52	2.39
±SD	1.12	1.45	1.30	0.99	0.84	0.82
n	13	35	49	53	51	54
Pyruvate, mmol/l						
Mean	0.13	0.12	0.11	0.10	0.09	0.10
±SD	0.06	0.05	0.04	0.03	0.03	0.03
n	12	31	37	41	43	44
Lactate/pyruvate ratio						
Mean	42.6	42.6	38.8	30.3	28.8	27.5
±SD	18.1	19.4	13.6	12.0	12.0	12.6
n	11	30	37	41	43	44

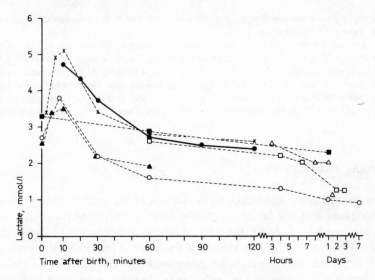

Fig. 1. Mean postnatal arterial lactate concentration in seven different investigations: ■ = BOSSART *et al.* [1]; ● = BRATTEBY and SWANSTRÖM; ▲ = DANIEL *et al.* [3]; ○ = KOCH and WENDEL [5]; △ = LAUTNER *et al.* [6]; □ = MATTHIEU *et al.* [8]; × = TUNELL *et al.* [10].

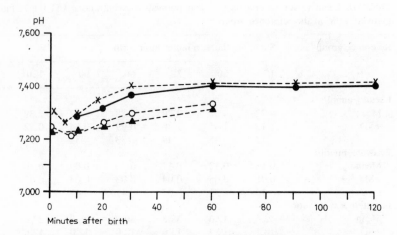

Fig. 2. Mean postnatal arterial pH in four investigations: ● = BRATTEBY and SWANSTRÖM; ▲ = DANIEL *et al.* [3]; ○ = KOCH and WENDEL [5]; × = TUNELL *et al.* [10].

The mean lactate values and standard deviations at different time intervals in the present study are very similar to those reported by TUNELL et al. [10]. They are also in agreement with results of MATTHIEU et al. [8], LAUTNER et al. [6], and BOSSART et al. [1], but are significantly higher than those of DANIEL et al. [3] and those of KOCH and WENDEL [5]. This difference is probably due to the effect of pH on lactate production. The pH values at different time intervals in the present study (fig. 2) are very similar to those of TUNELL et al. [10], but significantly higher than the mean pH in the infants of KOCH and WENDEL [5] and DANIEL et al. [3]. As a higher pH leads to an increased lactate production [2], at least part of the discrepancy in normal lactate values between these studies must be due to the difference in pH.

The mean pyruvate level of the normal infants in the present study is in agreement with that reported by MATTHIEU et al. [8] and LAUTNER et al. [6], but is approximately 0.1 mmol/l lower than the mean pyruvate level of normal newborn infants in the investigation of DANIEL et al. [3]. This discrepancy is probably due to methodological factors as the mean pyruvate value of normal adults in this latter investigation was also approximately 0.1 mmol/l higher than the normal adult pyruvate values obtained in our laboratory. The results on lactate/pyruvate ratio of normal infants in the present study are in accordance [6] or disagreement [3] with investigations presented earlier depending on the respective lactate or pyruvate values referred to above.

The mean lactate concentrations in asphyxiated infants, given in table IV, are higher than those in the normal infants. This difference is significant for samples taken later than 10 min after birth. The same is true for the pyruvate concentrations. The lactate/pyruvate ratio, however, did not differ significantly between the asphyxiated and normal infants at any sampling time.

A strong correlation (fig. 3) was found between lactate concentration and base deficit in blood samples obtained 30 min after birth (asphyxia group $r = 0.91$, $p < 0.001$; combined asphyxia + reference group $r = 0.69$, $p < 0.001$). The pyruvate concentration and base deficit were also significantly correlated in the asphyxia group ($r = 0.76$, $p < 0.01$) and in the combined reference and asphyxia group ($r = 0.59$, $p < 0.001$), while the lactate/pyruvate ratio showed no significant correlation to base deficit in any group (fig. 4).

The cumulative Apgar score (table V) was significantly correlated to the lactate concentration, and also to base deficit. A significant correlation between pyruvate concentration and cumulative Apgar score was found only in the combined asphyxia and reference group but not in the asphyxia group. No significant correlation was found between the

Table IV. Mean (±SD) arterial lactate and pyruvate concentration and the lactate/pyruvate ratio in the asphyxia group. At each sampling time, the significance of the differences from the mean values in the reference group (table III) were calculated by Student's unpaired *t*-test

Asphyxia group	Sampling times, minutes after birth					
	10	20	30	60	90	120
Lactate, mmol/l						
Mean	7.54	6.80	6.38	5.13	4.14	3.74
±SD	5.03	4.43	4.27	4.19	4.23	3.60
n	8	7	13	13	12	13
Differences from mean values in the reference group, p	n.s	0.01	0.001	0.001	0.02	0.02
Pyruvate, mmol/l						
Mean	0.21	0.17	0.19	0.17	0.16	0.15
±SD	0.11	0.03	0.15	0.11	0.11	0.11
n	7	5	11	11	11	11
Differences from mean values in the reference group, p	n.s.	0.05	0.01	0.001	0.001	0.01
Lactate/pyruvate ratio						
Mean	40.9	35.5	40.5	30.3	23.8	22.7
±SD	15.0	15.1	17.0	11.1	8.3	7.6
n	7	5	11	11	11	10
Differences from mean values in the reference group, p	n.s.	n.s.	n.s.	n.s.	n.s.	n.s.

lactate/pyruvate ratio and this cumulative Apgar score either in the asphyxia group or in the combined group.

The findings concerning the correlation between lactate concentration and base deficit and also between lactate concentration and cumulative Apgar score are in agreement with those of MATTHIEU *et al.* [8]. The observation that the postnatal pyruvate concentration has a positive but less strong correlation than lactate to different signs of fetal hypoxia is in accordance with earlier reports [9], but has not previously been described statistically. The use of the lactate/pyruvate ratio in the neonate as a biochemical indicator of past hypoxia has been critized [9], but a higher ratio has been observed in severely depressed infants than in normal newborn infants [3, 9].

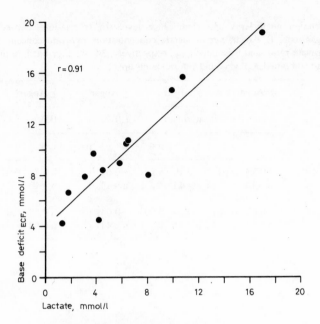

Fig. 3. Correlation between base deficit$_{ECF}$ and arterial lactate concentration in the asphyxia group 30 min after birth.

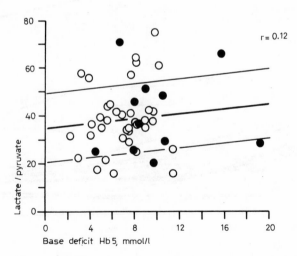

Fig. 4. Correlation between lactate/pyruvate ratio and base deficit in normal (○) and asphyxiated (●) infants.

Table V. Correlation coefficients and significance levels of correlations between cumulative Apgar score ($\sum$ Apgar) versus lactate concentration, pyruvate concentration, lactate/pyruvate ratio, and base deficit$_{ECF}$, respectively, 30 min after birth, in the asphyxia and the combined asphyxia and reference group

	$\sum$ Apgar lactate		$\sum$ Apgar pyruvate		$\sum$ Apgar L/P ratio		$\sum$ Apgar BD$_{ECF}$	
	r	p<	r	p<	r	p<	r	p<
Asphyxia group	−0.73 n = 13	0.01	−0.49 n = 11	n.s.	−0.04 n = 11	n.s.	−0.71 n = 13	0.01
Asphyxia and reference group	−0.62 n = 54	0.001	−0.50 n = 41	0.001	−0.04 n = 41	n.s.	−0.59 n = 54	0.001

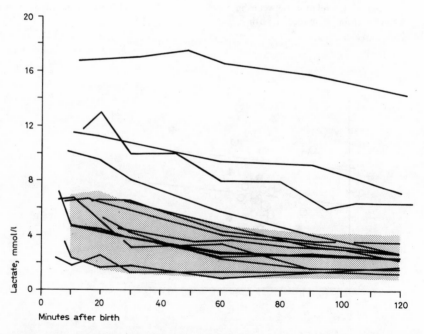

Fig. 5. Arterial lactate concentration in individual asphyxiated infants. The shaded area represents the mean ±2 SD of the reference group.

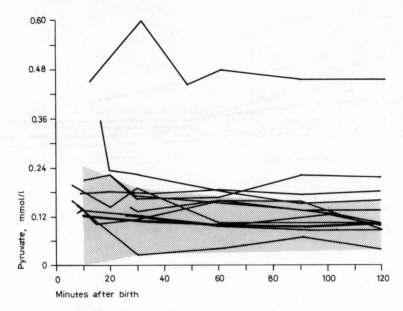

Fig. 6. Arterial pyruvate concentration in individual asphyxiated infants. The shaded area represents the mean ±2 SD of the reference group.

Considering the individual values of the asphyxiated infants and taking the mean + 2 standard deviations of the reference value as an upper limit of normality, only 4 infants had lactate values outside the normal range (fig. 5), but these were the infants with the most alarming signs of intrauterine distress. The lactate values for the individual infant followed the same percentile and kept at a fairly constant distance to the mean reference value throughout the observation period. For pyruvate also, 4 of the most asphyxiated infants had values outside the normal range during the first 2 h of life (fig. 6). The pyruvate concentrations in these 4 infants, however, showed a greater fluctuation than the corresponding lactate concentrations and in only 1 of the infants were all pyruvate values outside the normal range during the whole observation period. For the lactate/pyruvate ratio (fig. 7), all values for the asphyxiated infants lay within the normal range.

It has been argued [8] that the lactate concentration could be used as a quantitative measure of hypoxia in newborn infants. From the correlation between clinical signs of hypoxia and lactate level, this assumption would seem probable. The fact that only 4 of the 13 asphyxiated infants in the present study had lactate concentrations outside the normal range

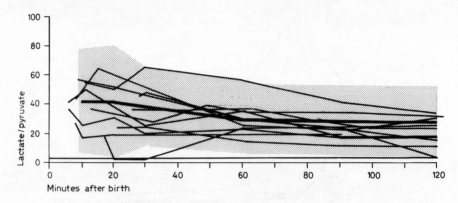

Fig. 7. Arterial lactate/pyruvate ratio in individual asphyxiated infants. The shaded area represents the mean ±2 SD of the reference group.

is therefore surprising and indicates that the use of postnatal lactate values for discrimination of prenatal hypoxia in the individual infant is limited in clinical practice. These 4 infants were more easily identified by their alarming clinical signs than by lactate analysis. What is the explanation for this poor discrimination of individual infants with prenatal hypoxia? Two main factors interfere with the postnatal biochemical assessment of prenatal hypoxia: (1) The nature of the postnatal normal values that have to be used as a reference, and (2) the influence of the timing of the prenatal hypoxia on the postnatal values.

The Influence of the Normal Values. Biochemical assessment of hypoxia from the lactate concentration, for example, is essentially more complicated postnatally than later in life, as the hypoxia is not estimated against a constant, narrow normal range. On the contrary, the normal range is wide and shows rapid changes early after birth.

The Effect of the Time Interval between Prenatal Hypoxia and Sampling. The postnatal lactate concentration after a given fetal hypoxia is largely dependent on the time interval between the hypoxia and birth. This means that a given fetal hypoxia occurring immediately before birth is more easily detected by postnatal lactate values than a hypoxic event occurring in the earlier part of delivery. This is illustrated by figure 8,

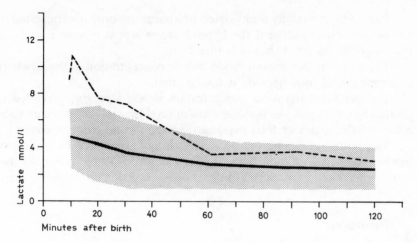

Fig. 8. Arterial lactate concentration in a newborn infant after a breech delivery with a short period of prenatal hypoxia (- - - -). 1-min Apgar score = 9. The shaded area represents the mean ±2 SD of the reference group.

which shows the lactate concentration in an infant with a breech delivery. The fetal heart rate pattern was normal and there were no other signs of fetal distress in this infant until the very last minute before birth, when there was a very short, intense hypoxic period. The infant recovered immediately after birth. The Apgar scores were normal, but the lactate, pH, and base deficit values early after birth were outside the normal range. Had the short period of hypoxia in this infant occurred 20 min earlier, it would probably not have been detectable by the postnatal lactate values.

Conclusions

The following conclusions are based on a group of infants fulfilling very strict criteria of normality and a group of infants with a clinically well-documented asphyxia, who had an adequate gas exchange and a rapid clinical recovery after resuscitation.

During the first 2 h of life, the normal lactate concentration is higher than the normal lactate concentration at rest later in life, and shows rapid changes and a wide range of variation.

The postnatal lactate concentration is closely correlated to postnatal clinical and biochemical signs of fetal hypoxia and discriminates a group of asphyxiated newborn infants from a group of normal infants.

Individual prenatally asphyxiated newborns are only discriminated by the lactate concentration if the hypoxic stress was very severe or if the fetal hypoxia occurred shortly before birth.

The value of the arterial blood lactate concentration in the newborn as a measure of fetal hypoxia is thus limited.

The postnatal pyruvate concentration is less strongly correlated to postnatal clinical and biochemical signs of fetal hypoxia and appears to be a less reliable index of fetal hypoxia than the lactate concentration.

The postnatal lactate/pyruvate ratio is not correlated to clinical or biochemical signs of fetal hypoxia and cannot serve to distinguish between asphyxiated and normal infants either individually or as a group.

References

1 BOSSART, H.: NIEDERHAUSERN, F., REY, I. et WEIHS, D.: pH sanguin, glycémie et lactacidémie chez la mère et l'enfant pendant et après l'accouchement normal. Gynecologia *165:* 146–151 (1968).
2 CAIN, S. M.: pH effects on lactate and excess lactate in relation to O_2 deficit in hypoxic dogs. J. appl. Physiol. *42:* 44–49 (1977).
3 DANIEL, S. S.; ADAMSONS, K., jr., and JAMES, L. S.: Lactate and pyruvate as an index of prenatal oxygen deprivation. Pediatrics *37:* 942–953 (1966).
4 GRIFFITHS, R.: The abilities of babies (London Press, London 1954).
5 KOCH, G. and WENDEL, H.: Adjustment of arterial blood gases and acid base balance in the normal newborn infant during the first week of life. Biol. Neonate *12:* 136–161 (1968).
6 LAUTNER, H.; KAISER, D. und WERNER, E.: Der Verlauf der Lactat- und Pyruvatkonzentration und des Lactat-Pyruvat-Quotienten in den ersten 48 Lebensstunden. Mschr. Kinderheilk. *116:* 257–259 (1968).
7 MARBACH, E. P. and WEIL, M. H.: Rapid enzymatic measurement of blood lactate and pyruvate. Use and significance of metaphosphoric acid as a common precipitant. Clin. Chem. *13:* 314–325 (1967).
8 MATTHIEU, J.-M.; GAUTIER, E.; PROD'HOM, L.-S. et FREI, L.: Acidose lactique et asphyxie de nouveau-né. Helv. paediat. Acta suppl. XXVI, pp. 3–27 (1971).
9 STAVE, U.: Metabolic effects in hypoxia neonatorum; in STAVE Physiology of the perinatal period, vol. 2, pp. 1043–1088 (Appleton Century Crofts, New York 1970).
10 TUNELL, R.; COPHER, D., and PERSSON, B.: The pulmonary gas exchange and blood gas changes in connection with birth; in STETSON and SWYER Neonatal intensive care, pp. 89–106 (Green, St. Louis 1976).

L.-E. BRATTEBY, MD, Perinatal Research Unit, University Hospital, *S-750 14 Uppsala* (Sweden)

Medicine and Surgery

Lactate in Acute Conditions. Int. Symp., Basel 1978, pp. 69–82 (Karger, Basel 1979)

Lactate in Acute Circulatory Failure

C. PERRET and J.-F. ENRICO

Intensive Care Unit and Respiratory Unit, Department of Medicine,
University Hospital, Lausanne

Metabolic acidosis during prolonged shock is a well recognized and documented fact. As early as 1918, CANNON [9] reported a relationship between the severity of shock and the decrease of pH in experimental animals. JERVELL [20], in 1928, and COURNAND et al. [12], in 1943, showed that metabolic acidosis of shock was bound to lactate accumulation in extracellular fluid.

Numerous studies carried out in the different types of shock confirmed these preliminary observations [6, 17, 19, 21, 25, 26, 29, 34, 39]. The common mechanism was considered to be an imbalance between tissue oxygen supply and metabolic needs, a condition later described by COHEN and WOODS [10] as the type A variety of lactic acidosis.

In 1964, PERETZ et al. [24] showed that blood lactate level was a useful prognostic index. There was a sharp increase in mortality rate when the initial arterial blood lactate concentration rose over 4.5 mmol/l. The same year, BRODER and WEIL [7] reported similar results using excess-lactate (XL) level which according to HUCKABEE [18] was supposed to be an indirect measurement of oxygen debt. Probability of survival was poor when the XL level exceeded 3 mmol/l and death was the rule, whatever the treatment, when it exceeded 4 mmol/l [7]. However, several objections were formulated, especially in endotoxin shock where no correlation could be demonstrated between the measured oxygen deficit and the lactate level [32, 33].

The present study was undertaken to evaluate the acid-base status in shock, to determine which factors other than anoxia could interfere with lactate metabolism and, finally, to test the prognostic value of hyperlactatemia in acute circulatory failure.

Patients and Methods

A group of 84 patients was studied, including 31 cases of hypovolemic shock, 25 cases of cardiogenic shock and 28 cases of septic shock. Diagnosis of shock was established in the presence of the following criteria: systolic arterial pressure below 90 mm Hg, a urine output less than 25 ml/h associated with peripheral sweating, cyanosis of the extremities and possible mental confusion. Arterial blood samples were collected in all the patients upon admission and repeated regularly along the evolution. Lactate and pyruvate were determined by the enzymatic method as described by HOHORST [16] and BÜCHER et al. [8].

Normal lactate value for our laboratory is 0.97 mmol/1 (±0.08) and for pyruvate 0.097 mmol/1 (±0.008). It corresponds to a normal L/P ratio of 10.

Results

Individual values for H+ concentration and PCO_2 obtained upon admission are presented in figure 1. It appears clearly that there is no evident relationship between the etiology of shock and the type of acid-base disorder observed. Upon admission, 43 patients were in chronic

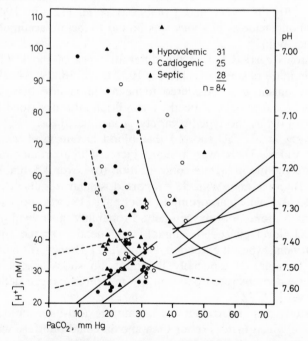

Fig. 1. Acid-base status measured upon admission in 84 shock patients. Individual values are reported on a H+/PCO_2 diagram.

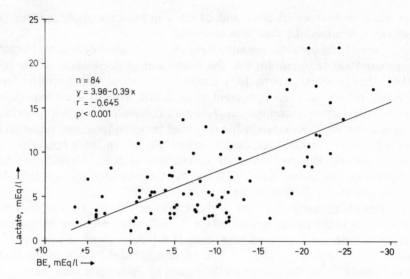

Fig. 2. Relation between arterial blood lactate concentration and base-excess value measured upon admission in 84 shock patients.

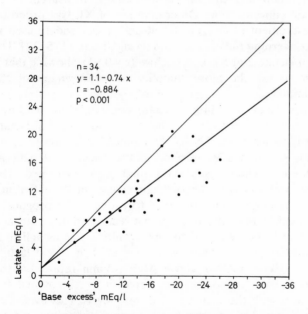

Fig. 3. Relation between arterial blood lactate concentration and base-excess value in 34 patients resuscitated for cardiac arrest.

or acute respiratory alkalosis and 24 cases in pure metabolic acidosis. In 17 cases, a mixed disorder was observed.

Correlation between negative base excess (or base deficit) and lactate concentration is significant but the coefficient of correlation is low ($r = 0.65$) (fig. 2). Some patients have a high-lactate value with a positive base excess; others show an important base deficit without a corresponding increase in lactate concentration. This poor correlation is rather surprising considering the close relationship observed between these two parameters in patients resuscitated for cardiac arrest (fig. 3). In these conditions of acute anoxia, the coefficient of correlation is 0.88. Considering the position of the line of equality, it can be assumed that protons released come essentially from lactic acid.

The discrepancy observed in shock between these two variables has potentially many causes among which the following might be considered: firstly the presence of acid-base disorders independent of anoxia; secondly, the potentially different fate of protons and lactate in the body fluids, and thirdly, the release by the tissues of acids other than lactic acid, as shown by increased levels of phosphate, citrate, 3-OH-butyrate and acetoacetate.

It may thus be concluded that neither the determination of base deficit nor that of anion gap can be substituted for measurement of arterial blood lactate concentration. Determination of XL was proposed as an indirect measurement of oxygen deficit and it was widely used in spite of criticisms concerning the concept and its significance [15, 22]. The correlation between lactate and XL is very close ($r = 0.95$), showing that it is of no advantage to use the more complicated measurement of XL which needs the simultaneous determination of pyruvate concentration.

Individual values of arterial blood lactate concentration measured upon admission are reported in figure 4. There is a wide range in lactate concentration which varies from 1.8 to 22.5 mmol/l. The lower mean value in cardiogenic shock can be ascribed to the selection of patients which excluded all cases where resuscitation had been performed. The highest mean value is observed in hemorrhagic shock but the difference from septic shock is not significant. In order to estimate the prognostic value of arterial blood lactate, the relation between mortality rate and maximal blood lactate level measured during evolution was established. Patients were separated into 7 groups according to their blood lactate concentration (fig. 5). The number above each column represents the number of patients in each group. On the whole, it is evident that the survival rate decreases with the increase of lactate levels. However, two facts can be pointed out. On the one hand, some patients had a fatal outcome with surprisingly low lactate levels. On the other hand, several

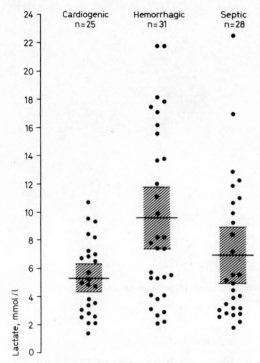

Fig. 4. Distribution of arterial blood lactate concentrations measured upon admission in 84 patients with shock. The shaded area corresponds to 2 SE.

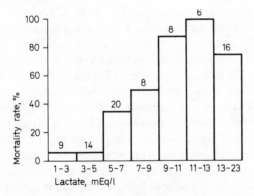

Fig. 5. Mortality rate as a function of the maximal arterial blood lactate level measured during the evolution in 81 shock patients.

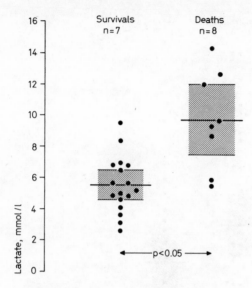

Fig. 6. Distribution of maximal arterial blood lactate concentrations in survivors and non-survivors from cardiogenic shock.

patients survived, with lactate blood levels superior to 13 mmol/l. Such a discrepancy suggests that lactatemia depends not only on the intensity and duration of tissue anoxia. Other factors might be implicated which could interfere with lactate metabolism and, by this way, diminish its prognostic value.

Now, let us consider the prognostic value of arterial blood lactate concentration in the different types of shock. Figure 6 refers to cardiogenic shock. Individual maximal lactate values have been separated in two groups according to the evolution. In the group of survivors, the mean lactate concentration is 5.6 mmol/l, a significantly lower value than that obtained in patients with a fatal outcome. However, there is an important overlap of individual values and patients with lactate concentration as high as 10 mmol/l may have a surprisingly favorable evolution. It is interesting to note that those patients who survived with high-lactate levels were in acute anoxia and were promptly improved by the treatment. Similar conditions are observed in patients who have been successfully resuscitated. Blood lactate evolution in 8 patients who recovered from a cardiac arrest in spite of an impressive hyperlactatemia is reported in figure 7. Comparable values were reported after 'grand mal' seizures [30]. These facts are now widely accepted. They demonstrate that very

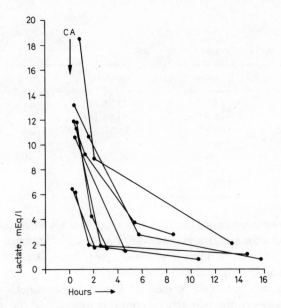

Fig. 7. Lactate removal after cardiac arrest (CA) in 8 patients successfully resuscitated.

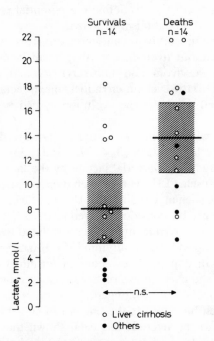

Fig. 8. Distribution of maximal arterial blood lactate concentrations in survivors and non-survivors from hemorrhagic shock.

Table I. Haemorrhagic shock. Maximal lactate values

	Survivals	Deaths	p
Total patients	7.98	13.86	n.s.
(n = 28)	(n = 14)	(n = 14)	
Cirrhosis	10.53	15.57	n.s.
(n = 18)	(n = 9)	(n = 9)	
Others	3.39	10.77	<0.05
(n = 10)	(n = 5)	(n = 5)	
p	<0.02	n.s.	

high lactate levels do not necessarily indicate a poor prognosis when they are related to acute and transient anoxia, a condition somewhat comparable to that of maximal exercise. Conversely, a less severe hyperlactatemia implies a poor prognosis if progressive and prolonged.

When considering mean lactate values in hemorrhagic shock, no significant difference between the two groups could be established (fig. 8). Of particular interest was the favorable evolution of several patients with liver cirrhosis in spite of extremely high blood lactate levels. So it appeared of major importance for prediction to define the potential role of preexisting liver disease on lactate metabolism in anoxia.

Mean maximal lactate values in survivors and non-survivors are reported in table I. In patients without liver disease, that mean lactate level is significantly different in survivors and non-survivors and so represents a good prognostic index. However, in cirrhotics, mean lactate concentration is strikingly increased whatever the evolution and so is of no prognostic value.

These data are in agreement with the previous studies showing the predominant role of liver in lactate homeostasis [3, 4, 27, 28, 31, 35–38, 40]. In normal conditions, net extraction of arterial lactate by the kidney [42], the heart [23], and skeletal muscle [13] has also been demonstrated, although less effectively. In the particular condition of shock, it can be inferred that kidney extraction must be considerably reduced with the early development of renal ischemia. The capacity of the rat isolated liver to metabolize lactate is very high, up to 2.0 mmol/g/min [35]. BALLINGER et al. [2] in experimental studies in dogs gave evidence of an effective removal of lactate by the liver in the early phase of hemorrhagic shock. However, acute circulatory failure, when prolonged, goes on with a decreased hepatic blood flow which both lowers the amount of substrate and the oxygen supply delivered to the liver. It has been shown that a severe liver ischemia is necessary to decrease liver capacity to metabolize lactate [38], but it seems probable that in patients with previous hepatic

Table II. Blood lactate concentration at different sampling sites in a 70-year-old cirrhotic patient with a septic shock

	Lactate mmol/l
Femoral artery	5.2
Pulmonary artery	5.15
Superior vena cava	5.0
Hepatic vein	5.6

disease a moderate ischemia is able to greatly impair lactate uptake and so to contribute to raise lactate levels. The data reported on table II refer to a cirrhotic patient with a septic shock. Paradoxically, the lactate concentration is highest in the hepatic vein suggesting a failure of liver lactate removal, or even a switch to production. These results suggest the following conclusion: a high-lactate level does not necessarily indicate a poor prognosis in patients with previous liver dysfunction. In these conditions, because of the lowered capacity of clearing lactate and, possibly, a decreased tolerance to anoxia, a reversible oxygen deficit may lead to a severe hyperlactatemia.

Analysis of our data in septic shock shows that no patient survived with a maximal lactate level superior to 7 mmol/l while, on the other hand, some patients died with lactate levels as low as 2 mmol/l (fig. 9). In order to test the possible role of anoxic liver dysfunction on the maximal lactate value observed during evolution, patients were separated into two groups (table III). In the first group are collected patients who developed biological signs of extensive hepatic cytolysis, confirmed histologically in those with a fatal evolution. In the second group, liver function tests and liver histology, when performed, were maintained in normal limits.

It is interesting to note that here too the development of liver dysfunction secondary to prolonged ischemia contributes in increasing arterial blood lactate concentration. First, in survivors the mean of the maximal lactate values obtained in patients with liver dysfunction is twice that measured in cases with normal liver function tests. The difference is still larger in non-survivors with respective values of 12.8 and 4.7 mmol/l. The difference is highly significant. Secondly, mean lactate value in survivors with liver cytolysis is not significantly different from that of non-survivors without liver dysfunction. Thirdly, in this latter group, lactate level is surprisingly low compared with that measured in patients dying from other types of shock.

These observations are in agreement with those reported by ROSEN-

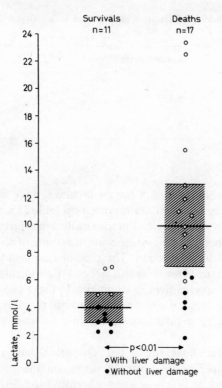

Fig. 9. Distribution of maximal arterial blood lactate concentrations in survivors and non-survivors from septic shock.

Table III. Septic shock. Maximal lactate values

	Survivals	Deaths	p
Total patients	4.02	9.96	<0.01
(n = 28)	(n = 11)	(n = 17)	
With liver damage	5.89	12.84	<0.02
(n = 15)	(n = 4)	(n = 11)	
Without liver damage	2.95	4.67	n.s.
(n = 13)	(n = 7)	(n = 6)	
p	<0.05	<0.01	

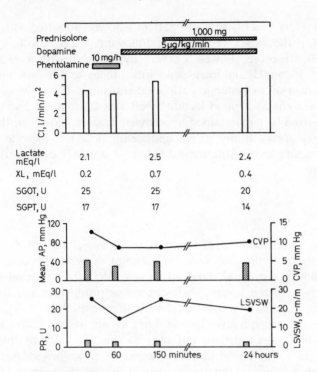

Fig. 10. Hemodynamic evolution in a 66-year-old patient admitted for a septic shock. The high cardiac index (CI) goes on with a fall in arterial blood pressure (AP) and systemic peripheral resistance (PR). Note the persistently low arterial blood lactate concentration.

BERG and RUSH [32, 33] who demonstrated that blood lactate levels in dogs were significantly lower in endotoxic shock than in hypovolemic shock, although the estimated cumulative oxygen debt was similar in the two conditions. The authors suggested the possible role of a tissue metabolism inhibition secondary to the action of bacterial endotoxin. Another interpretation was proposed, considering the particular 'hyperdynamic state' goes on with a particular distribution of organs' perfusion. WYLER et al. [41] demonstrated in the unanesthetized monkey decreased total peripheral resistances [1, 5, 11, 14, 21]. This so-called 'hyperdynamic state' goes on with a particular distribution of organs' perfusion. WYLER et al. [41] demonstrated in the unanesthetized monkey that liver blood flow was increased in the early phase of shock. This contributes to maintain adequate oxygen supply to hepatic cells and to realize proper conditions for an effective removal of lactate. This would explain the paradoxically low lactate levels in patients dying of septic shock. Figure 10 refers to a 66-year-old patient who was admitted within a few hours after the onset of a shock following a prostatectomy. Hemodynamic investigations showed a severe hypotension with a mean blood pressure of 40 mm Hg, an increased cardiac index to 4.5 l/min and a considerable reduction in systemic peripheral resistance. There was no sign of hepatic cytolysis and blood lactate levels was 2.1 mmol/l. Evolution was characterized by a maintained hyperdynamic state. However, the patient's condition grew rapidly worse and death occurred, despite a persistently low lactate level. Autopsy did not show optically demonstrable hepatic lesions.

Conclusions

In conclusion, determination of base excess or anion gap does not permit one to predict lactate concentration, except in particular conditions characterized by acute and transient anoxia. Blood lactate level in shock is determined by the balance between tissue production and the overall capacity to clear it. The liver appears as the most effective organ to remove lactate. Consequently, lactate level is not only related to cellular anoxia and oxygen deficit, but also to the adequacy of liver function. Liver failure may be the direct consequence of prolonged shock or it may result from previous hepatic damage, as in cirrhotic patients. In these conditions, severe hyperlactatemia may develop in the absence of irreversible tissue anoxia. Conversely, when liver perfusion is preserved, as possibly occurs in certain cases of septic shock, hepatic removal of lactate is supposed to be sufficiently effective to maintain low blood lactate levels despite severe tissue anoxia. In consequence, the significance of hyperlactatemia during shock is complex and prediction of the outcome from an isolated value is necessarily hazardous. A better estimation of prognosis can be expected when following the evolution of lactatemia during treatment. Recent technical facilities make it possible to monitor blood lactate concentration. This constitutes an essential step in estimating the adequacy of the treatment.

References

1   ALBRECHT, M. and CLOVES, G. H. A.: The increase of circulatory requirements in the presence of inflammation. Surgery, St Louis 56: 158–162 (1964).
2   BALLINGER, W. F., II.; VOLLENWEIDER, H., and MONTGOMERY, E. H.: The response of canine liver to anaerobic metabolism induced by hemorrhagic shock. Surgery Gynec. Obstet. 112: 19–26 (1961).

3   BERRY, M. N. and SCHEUER, J.: Splanchnic lactic acid metabolism in hyperventilation, metabolic alkalosis and shock. Metabolism 16: 537–547 (1967).
4   BERRY, M. N.: The liver and lactic acidosis. Proc. R. Soc. Med. 60: 1260–1262 (1967).
5   BLAIN, C.; ANDERSON, T. O.; PIETRAS, R. J., and GUNNAR, R. M.: Immediate hemodynamic effects of Gram-Negative vs Gram-positive bacteremia in man. Archs intern. Med. 126: 260–264 (1970).
6   BLAIR, E.; MCLAUGHLIN, J. S.; COWLEY, R. A., and TAIT, M. K.: Lactacidemia in clinical shock. Clin. Res. 13: 319–323 (1965).
7   BRODER, G. and WEIL, M. H.: Excess lactate: an index of reversibility of shock in human patients. Science 143: 1457–1459 (1944).
8   BUCHER, T. H.; CZOK, R.; LAMPRECHT, W. und LATZKO, E.: Pyruvat; in BERGMEYER Methoden der enzymatischen Analyse, pp. 253–265 (Verlag Chemie, Weinheim 1962).
9   CANNON, W. B.: Acidosis in shock. Bull. med. Paris i: 424 (1918); cited by COHEN, R. D. and WOODS, H. F.: Clinical and biochemical aspects of lactic acidosis, p. 8 (Blackwell, Oxford 1976).
10   COHEN, R. D. and WOODS, H. F.: Clinical and biochemical aspects of lactic acidosis, pp. 41–42 (Blackwell, Oxford 1976).
11   COHN, J. D.; GREENSPAN, M.; GOLDSTEIN, C. R.; GUDWIN, A. L.; SIEGEL, J. H., and GUERCIO, R. M. DEL: Arteriovenous shunting in high cardiac output shock syndromes. Surgery Gynec. Obstet. 127: 282–291 (1968).
12   COURNAND, A.; RILEY, R. L.; BRADLEY, S. E.; BREED, E. S.; NOBLE, R. P.; LAUSON, H. D.; GREGERSEN, M. I., and RICHARDS, D. W.: Studies of the circulation in clinical shock. Surgery 13: 964–995 (1943).
13   DUNN, R. B. and CRITZ, J. B.: Uptake of lactate by dog skeletal muscle in vivo and the effect of free fatty acids. Am. J. Physiol. 229: 255–259 (1975).
14   ENRICO, J.-F.; HADORN, R.; POLI, S. et PERRET, C.: Intérêt de l'exploration hémodynamique au cours du choc septique; in Journées de Réanimation de l'Hôpital Claude Bernard, pp. 123–148 (Librairie Arnette, Paris 1971).
15   HARRIS, P.; BATEMAN, M., and GLOSTER, J.: Relation between the cardiorespiratory effects of exercise and the arterial concentration of lactate and pyruvate in patients with rheumatic heart disease. Clin. Sci. 23: 531–544 (1962).
16   HOHORST, H.-J.: Laktat; in BERGMEYER Methoden der enzymatischen Analyse, pp. 266–270 (Verlag Chemie, Weinhein 1962).
17   HOPKINS, R. W.; SABGA, G.; PENN, I., and SIMEONE, F. A.: Hemodynamic aspect of hemorrhagic and septic shock. J. Am. med. Ass. 191: 731–735 (1965).
18   HUCKABEE, W. E.: Relationships of pyruvate and lactate during anaerobic metabolism. II. Exercise and formation of O_2 debt. J. clin. Invest. 37: 255–263 (1958).
19   HUCKABEE, W. E.: Abnormal resting blood lactate. I. Significance of hyperlactatemia in hospitalized patients. Am. J. Med. 30: 833–839 (1961).
20   JERVELL, O.: Investigation of the concentration of lactic acid in blood and urine under physiologic and pathologic conditions. Acta med. scand. Suppl. 24 (1928).
21   MCLEAN, L. D.; MULLIGAN, W. C.; MCLEAN, A. P. H., and DUFF, J. H.: Patterns of septic shock in man. A detailed study of 56 patients. Ann. Surg. 166: 543–562 (1967).
22   OLSON, R. E.: 'Excess lactate' and anaerobiosis. Ann. intern. Med. 59: 960–963 (1963).
23   OPIE, L. H.: Metabolism of the heart in health and disease. Part I. Am. Heart J. 76: 685–698 (1968).
24   PERETZ, D. I.; MCGREGOR, M., and DOSSETOR, J. B.: Lactic acidosis: a clinically significant aspect of shock. Can. med. Ass. J. 90: 673–675 (1964).
25   PERETZ, D. I.; SCOTT, H. M.; DUFF, J.; DOSSETOR, J. B.; MCLEAN, L. D., and

McGREGOR, M.: The significance of lactacidemia in the shock syndrome. Ann. N.Y. Acad. Sci. *119:* 1133–1141 (1965).

26 PERRET, C.: L'acidose métabolique au cours du choc; in Journées de réanimation neuro-respiratoire, Hôpital Claude Bernard 1966, pp. 129–138 (Librairie Arnette, Paris 1966).

27 PERRET, C.; ENRICO, J.-F.; MONTANI, S. et PAPPALARDO, G.: Le rôle de la nécrose hépatique dans la pathogenèse de l'acidose lactique 'spontanée'. Schweiz. med. Wschr. *97:* 666–671 (1967).

28 PERRET, C.; POLI, S., and ENRICO, J.-F.: Lactic acidosis and liver damage. Helv. med. Acta *35:* 377–405 (1969/70).

29 PHILLIPSON, E. A. and SPROULE, B. J.: The clinical significance of elevated blood lactate. Can. med. Ass. J. *92:* 1334–1338 (1965).

30 POLI, S.; KALBERMATTEN, J. P. DE; ENRICO, J.-F. et PERRET, C.: Crise éileptique et perturbations acido-basiques (abstr.). Helv. med. Acta Suppl. *50:* 120 (1970).

31 ROWELL, L. B.; KRAINING, K. K., II; EVANS, T. O.; KENNEDY, J. W.; BLACKMON, J. R., and KUSUMI, F.: Splanchnic removal of lactate and pyruvate during prolonged exercise in man. J. appl. Physiol. *21:* 1773–1783 (1966).

32 ROSENBERG, J. C. and RUSH, B. F.: Lethal endotoxin shock. Oxygen deficit, lactic acid levels and other metabolic changes. J. Am. med. Ass. *196:* 767–769 (1966).

33 ROSENBERG, J. C. and RUSH, B. F.: Blood lactic acid levels in irreversible and lethal endotoxin shock. Surgery Gynec. Obstet. *126:* 1247–1250 (1968).

34 RUSH, B. F., jr.; ROSENBERG, J. C., and SPENCER, F. C.: Changes in oxygen consumption in shock. Correlation with other known parameters. J. surg. Res. *5:* 252–255 (1965).

35 SCHIMASSEK, H.: Der Einfluss der Leber auf den extracellulären Redox-Quotienten Lactat/Pyruvat. Versuche an der insoliert perfundierten Rattenleber. Biochem. Z. *336:* 468–473 (1963).

36 SCHRODER, R.; GUMPERT, J. R. W.; PLUTH, J. R.; ELTRINGHAM, W. K.; JENNY, M. E., and ZOLLINGER, R. M.: The role of the liver in the development of lactic acidosis in low flow states. Post-grad. med. J. *45:* 566–570 (1969).

37 SOFFER, L. J.; DANTES, D. A., and SOBOTKA, H.: Utilization of intravenously injected sodium *d*-lactate as a test of hepatic function. Archs intern. Med. *62:* 918–924 (1938).

38 TASHKIN, D. P.; GOLDSTEIN, P. J., and SIMMONS, D. H.: Hepatic lactate uptake during decreased liver perfusion and hypoxemia. Am. J. Physiol. *232:* 968–974 (1972).

39 UDHOJI, V. N. and WEIL, M. H.: Hemodynamic and metabolic studies on shock associated with bacteremia. Observations on 16 patients. Ann. intern. Med. *62:* 966–978 (1965).

40 WOODS, H. F. and KREBS, H. A.: Lactate production in the perfused rat liver. Biochem. J. *125:* 129–139 (1971)

41 WYLER, F.; FORSYTH, R. P.; NIES, A. S.; NEUTZE, J. M., and MELMON, K. L.: Endotoxin-induced regional circulatory changes in the unanesthetized monkey. Circulation Res. *24:* 777–786 (1969).

42 YUDKIN, J. and COHEN, R. D.: The contribution of the kidney to the removal of a lactic acid load in the conscious rat under normal and acidotic conditions. Clin. Sci. mol. Med. *48:* 121–131 (1975).

C. PERRET, MD, Intensive Care Unit and Respiratory Unit, Department of Medicine, University Hospital, *CH-1011 Lausanne* (Switzerland)

Lactate in Acute Conditions. Int. Symp., Basel 1978, pp. 83–101 (Karger, Basel 1979)

Lactate Metabolism in Diabetes

MALCOLM NATTRASS and K. G. M. M. ALBERTI

Faculty of Medicine, Chemical Pathology and Human Metabolism, South Laboratory and Pathology Block, General Hospital, Southampton

Introduction

The resurgence of interest in lactic acidosis over the past few years is due to a large extent to iatrogenic lactic acidosis occurring in patients with diabetes mellitus. The therapeutic use of biguanides in patients with maturity-onset (type 2) diabetes has resulted in more than 300 published cases of lactic acidosis [30]. This number is sufficiently large to deter enthusiastic reporting of single cases and it can be reasonably assumed that many more unreported cases and probably an equal number of unrecognized cases have occurred.

While few people would deny the causal role of biguanides in producing lactic acidosis, the relationship of diabetes mellitus *per se* and the occurrence of lactic acidosis is less clear. Does lactic acidosis occur more commonly in diabetics who are not receiving biguanides than in the normal population? – a suggestion that was made in some of the earlier reports of lactic acidosis.

Although lactic acidosis has been emphasized, there has been remarkably little attention paid to general aspects of lactate metabolism in diabetes, and in particular to the possible occurrence of hyperlactataemia without acidosis. It is apparent that insulin with its manifold actions must exert profound effects upon lactate production and utilization, and it is to be expected that iatrogenic abnormalities in lactate metabolism would occur with present insulin regimens where large quantities of insulin are deposited in subcutaneous tissue, and absorbed into the systemic circulation with little relation to the insulin needs of metabolism. It would indeed be remarkable if such crude therapy were to result in normal blood lactate concentrations and lactate metabolism, when in normal subjects minute-to-minute changes in insulin secretion into the portal system exert such a fine degree of metabolic control.

In this review, we shall briefly discuss lactate and pyruvate metabolism in normal man, then examine blood lactate behaviour in normal subjects and untreated type 2 diabetics. This will be followed by a discussion of the changes of lactate metabolism that occur with treatment.

Lactate Metabolism in Normal Man

Lactate turnover amounts to some 140 g/day in normal man. About half of this is derived from glucose, the rest coming from glycogen, alanine and other amino acids. Lactate is important as an end product of anaerobic metabolism and as a gluconeogenic precursor. Lactate is, however, produced from, and metabolized through, pyruvate. The equilibrium for the interconversion of lactate and pyruvate is such that quantitatively, lactate is of greater import; nevertheless, it is the production of pyruvate and its further metabolism that governs lactate concentrations in normal man.

A. The Role of Pyruvate (fig. 1)

Pyruvate occupies a central role in carbohydrate metabolism in man. Production of pyruvate from glucose or glycogen takes place in the cytoplasm of the cell and can occur anaerobically. Further oxidation of pyruvate takes place in the mitochondria and is oxygen dependent. The first stage in pyruvate oxidation is its irreversible conversion to acetyl-CoA under the influence of the enzyme complex pyruvate dehydrogenase. This acetyl-CoA can enter the tricarboxylic acid cycle or enter various pathways of lipid metabolism such as ketogenesis or fatty acid synthesis and be further oxidized. Pyruvate oxidation therefore represents a net loss of carbohydrate with concomitant energy production.

A second metabolic fate of pyruvate allows for the conservation of carbohydrate resources and hence the maintenance of blood glucose concentration by gluconeogenesis. Under appropriate conditions, pyruvate is carboxylated to oxaloacetate through the enzyme pyruvate carboxylase. Oxaloacetate can then be converted to phosphoenolpyruvate and hence to glucose. The interconversion of pyruvate and the amino acid alanine by the enzyme alanine aminotransferase allows alanine to function as the major gluconeogenic substrate derived from protein.

The metabolism of glucose to pyruvate, however, generates reduced nicotinamide adenine dinucleotide (NADH) from its oxidized form (NAD). Under resting aerobic conditions, NADH is oxidized intramitochondrially, but where this is impaired an alternative method of oxidation is necessary to restore the cytosolic pool of NAD necessary for

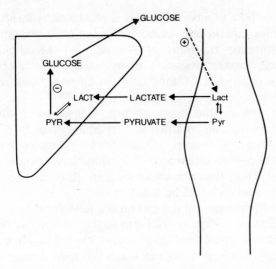

Fig. 1. The Cori cycle. Liver is portrayed on the left and muscle on the right. Lact = Lactate; Pyr = pyruvate; + and − indicate the effects of insulin upon the metabolic pathways, + = activation; − = inhibition. From [3b], with permission.

further glucose metabolism. Under such conditions, the conversion of pyruvate to lactate fulfills this role and lactate production becomes quantitatively more important than under resting, aerobic conditions.

B. The Regulation of Pyruvate Production

In the post-absorptive state blood glucose concentration in man is maintained by hepatic glycogenolysis and gluconeogenesis. After an overnight fast, 75% of glucose released by the liver is derived from glycogenolysis and 25% from gluconeogenesis [20]. Glycogen storage is promoted by insulin and the low concentrations of insulin in the fasted state aid maintenance of blood glucose concentration. Metabolism of glycogen to produce glucose is controlled mainly by the enzyme cascade of phosphorylase [23]. Phosphorylase activity is increased by glucagon (in liver) and adrenaline (liver and muscle) by increased conversion from its inactive to active form. Thus, increased secretion of these hormones promotes glycogen breakdown, glucose production and metabolism and increased pyruvate and lactate production.

The uptake of glucose by a number of tissues is regulated by insulin. This is particularly true for muscle cells where increasing concentrations of insulin promote membrane transport of glucose [28]. Membrane transport of glucose and further oxidation is inhibited by fatty-acid oxidation [37] and the regulation of fatty-acid production from tri-glycerides is a further step which is regulated by insulin.

The key regulatory step in glucose metabolism to pyruvate is the conversion of fructose-6-phosphate to fructose-1:6-diphosphate by the enzyme phosphofructokinase. This enzyme is regulated by a large number of factors including citrate concentration and ATP concentration both of which inhibit the enzyme. Thus, under conditions where citrate or energy production are impaired, glycolysis will be accelerated.

The net production of pyruvate will depend on the relative activity of the production and utilization pathways. Net production in muscle, for example, will increase if the stimulation of glycolysis quantitatively exceeds the stimulation of pyruvate dehydrogenase and this indeed appears to be the case. In addition, pyruvate is produced from carbon skeletons of amino acids, and the α-ketoacids, and insulin will tend to decrease production of these units by inhibiting intracellular proteolysis.

C. The Regulation of Pyruvate Utilization for Gluconeogenesis

The majority of enzymic steps in glycolysis are readily reversible and catalyzed by only one enzyme. The irreversible steps, however, necessitating alternative enzymes, are key regulatory steps for glycolysis and gluconeogenesis. The conversion of pyruvate to phosphoenolpyruvate is one such part of the pathway and involves intramitochondrial conversion of pyruvate to oxaloacetate (under the influence of pyruvate carboxylase) and then to phosphoenolpyruvate by phosphoenolpyruvate carboxykinase. Similarly, the enzyme fructose-1:6-diphosphatase is unique to gluconeogenesis and responsible for the conversion of fructose-1:6-diphosphate to fructose-6-phosphate.

Insulin inhibits gluconeogenesis. It is not clear whether this is due to inhibition of phosphoenolpyruvate carboxykinase via a decrease of cyclic 3'5'-AMP or by decreasing substrate supply for gluconeogenesis. The latter is linearly related to the rate of gluconeogenesis in the liver [16].

Glucagon may influence the rate of gluconeogenesis at both the pyruvate to phosphoenolpyruvate and fructose-1:6-diphosphate to fructose-6-phosphate steps. In addition, glucagon enhances hepatic extraction of the substrate alanine [31], thus promoting gluconeogenesis. Cortisol has similar effects, promoting the flow of extra gluconeogenic precursors from the periphery to the liver, and inducing key gluconeogenic enzymes [43].

D. The Regulation of Pyruvate Oxidation

The regulation of pyruvate dehydrogenase is achieved by a number of factors [15]. The enzyme complex exists in an active and inactive form and the conversion of one to the other is regulated enzymatically.

The ratio of acetyl-CoA to CoA is a major regulatory factor and the means by which fatty acid metabolism and ketogenesis exert an inhibitory effect upon pyruvate oxidation [35]. Increasing concentrations of pyruvate have been shown to stimulate production of the active form of pyruvate dehydrogenase [27].

Evidence that insulin affects the activity of pyruvate dehydrogenase arises from the observation that pyruvate stimulation of the enzyme is less marked in alloxan-dabetic rats [27]. In addition, COORE et al. [12] showed that insulin increased the proportion of active to inactive enzyme in adipose tissue with a direct action persisting in mitochondrial fractions of adipose tissue exposed to insulin.

In contrast, adrenaline inhibits pyruvate dehydrogenase activity. It is possible that both the insulin and adrenalin effects are mediated by changes in the intramitochondrial calcium concentration.

Lactate Metabolism in Diabetes

Insulin thus exerts profound effects upon both pyruvate production and utilization, and hence lactate production and utilization. These effects are both direct, e.g. by stimulating cellular uptake of glucose, and indirect, e.g. through effects on other hormones and metabolites. Thus, diabetes mellitus with absolute insulin deficiency or decreased insulin action relative to blood glucose concentration should be expected to result in a widespread derangement of lactate metabolism.

A. Blood Lactate Concentration in Normal Subjects

In normal subjects (fig. 2), blood lactate concentration throughout the day is maintained within narrow limits with maximum concentrations occurring after meals. The increased circulating concentrations of insulin at these times will both inhibit hepatic glucose production from lactate and stimulate lactate production by glycolysis. It is well recognized that the concentrations of insulin which produce this effect are vastly different [21, 38]. Thus, hepatic glucose production is inhibited by one tenth of the insulin concentration needed to promote peripheral glucose utilization. Regardless of the specific mechanism involved, the suggestion that insulin has an important role in determing blood lactate and pyruvate concentration is given weight by the finding that mean diurnal serum insulin concentration correlates closely with the mean pyruvate concentration [2].

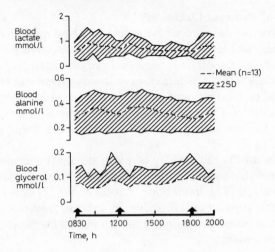

Fig. 2. Metabolic rhythms of blood lactate, alanine, and glycerol concentrations in 13 normal subjects. Data based on half-hourly blood sampling from 0830 to 2000 h. The dotted line indicates the mean and the shaded area mean ±2 standard deviations. The large arrows represent meals and the small arrows snacks.

B. Lactate Metabolism in Untreated Maturity-Onset Diabetics

DOAR and CRAMP [17] showed that subjects with untreated maturity-onset diabetes had an impairment of the rate of lactate removal compared with control subjects of equivalent weight. A considerable difference was present in mean age between their two groups; however, with control subjects having a mean age of 28.9 years compared with a mean age of 50.2 years in maturity-onset diabetics [17]. In view of the finding that blood lactate concentration is significantly higher in 'old' normal subjects than in young normal subjects [1], the difference observed by DOAR and CRAMP may be due to factors other than maturity-onset diabetes. Although a normal blood lactate concentration is usually found in maturity-onset diabetics, it should be noted that this does not necessarily imply normal lactate metabolism in that production and utilization could both be grossly disordered with a normal circulating concentration.

C. Blood Lactate Concentration in Maturity-Onset Diabetics Treated by a Carbohydrate-Restricted Diet

Figure 3 shows fasting, 2-hour post-breakfast, and 12-hour mean blood lactate concentration in normal subjects and in diabetics treated by dietary therapy only. The differences between the two groups of subjects are not significant. In 12-hour diurnal studies with half-hourly blood

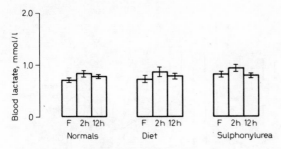

Fig. 3. Blood lactate concentration (mean ± SEM) in normal subjects, and maturity-onset diabetics treated by dietary restriction alone or with sulphonylurea therapy. F = Fasting; 2 h = 2 h post-breakfast: 12 h = mean concentration of half-hourly blood samples over 12 h.

sampling blood lactate concentration may be abnormal at certain times during the day if marked changes in blood glucose concentration occur. In addition, although blood lactate concentration may be within the normal range in these patients, the pattern seen may vary markedly from that of normal subjects.

D. Blood Lactate Concentration during Sulphonylurea Therapy

Blood lactate concentration in diabetics treated by sulphonylurea therapy is similar to that seen in normal subjects (fig. 3) although the fasting concentration of blood lactate is significantly higher during sulphonylurea therapy. The reason for this latter finding is obscure and turnover studies will be necessary to assess whether this is of biological as well as of statistical significance.

E. Blood Lactate Concentration during Biguanide Therapy

The hyperlactataemic effect of biguanide therapy was recognized shortly after the introduction of phenformin into anti-diabetic therapy [13, 19]. Only in recent years, however, with the escalating number of cases of lactic acidosis during biguanide therapy has a causal role been established. This has led to a withdrawal of phenformin from the drug market in a number of countries, leaving buformin and metformin as available therapies.

It is clear that increased circulating concentrations of lactate during biguanide therapy do not constitute an idiosyncratic response by the patient but are present to varying degrees in all patients taking biguanides.

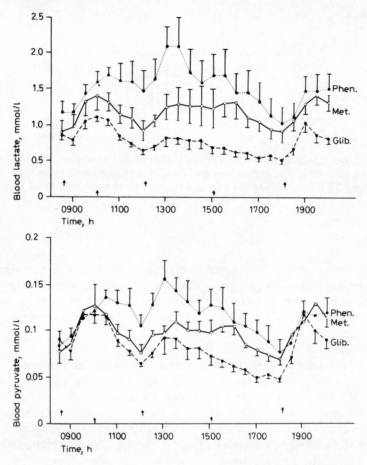

Fig. 4. Diurnal patterns of blood lactate and pyruvate concentrations in patients taking phenformin (Phen.), metformin (Met.) or glibenclamide (Glib.). Values are given as mean ± SEM. Arrows indicate meals and triangles snacks. Reprinted from NATTRASS *et al.* [32], by kind permission of the publishers.

We have studied several groups of biguanide-treated patients. These include patients treated by phenformin or metformin in combination with diet [32], patients with and without complications treated by sulphonylurea and phenformin therapy [33] and patients treated by sulphonylurea and metformin therapy [34]. Whenever either phenformin or metformin forms part of the therapy, blood lactate concentration is elevated above the normal range. Figure 4 shows the results obtained

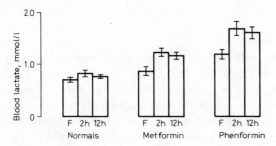

Fig. 5. Blood lactate concentration (mean ± SEM) in normal subjects and maturity-onset diabetics during metformin or phenformin therapy. Symbols as in figure 3.

during therapy with phenformin, metformin or glibenclamide in the same group of patients. Blood glucose concentration was not significantly different during phenformin or metformin therapy although it was lower during glibenclamide therapy. Metformin (1,500 mg daily) produced an exaggeration of the normal diurnal changes in blood lactate concentration while during phenformin therapy (100 mg daily) lactate levels were even more clearly abnormal. The pooled data from our studies is shown in figure 5. Fasting, 2-hour post-breakfast, and 12-hour mean concentrations are all significantly higher during either phenformin or metformin therapy. The difference between phenformin and metformin therapies is also significant ($p < 0.01$) for fasting, 2-hour post-breakfast, and 12-hour mean concentrations. It is of particular interest and of clinical relevance that the effects of metformin on blood lactate concentration, although clearly present, are consistently less than the effects of phenformin.

The hypoglycaemic effect of biguanides is produced by: (a) inhibition of glucose absorption in the small intestine; (b) increased peripheral utilization of glucose, and (c) inhibition of gluconeogenesis. Either of the last 2 mechanisms alone or in combination might be expected to produce hyperlactataemia and this effect therefore cannot be divorced from their hypoglycaemic effect. The subcellular action of biguanides has been shown by SCHÄFER [39] to be an alternation in the electro-static potential across the mitochondrial membrane leading to a more reduced state within the cytosol of the cell. This is reflected by an increased lactate/pyruvate ratio. Either as a consequence of this or in addition intramitochondrial oxidation of NADH appears to be impaired, resulting in a more reduced intramitochondrial state reflected in the 3-hydroxybutyrate/acetoacetate ratio.

Insulin

A. Diabetic Ketoacidosis

Several authors have estimated blood lactate concentrations in patients with diabetic ketoacidosis [24, 42]. Levels are on average greater than in stable diabetics, and in 15% of patients are above 5 mmol/l. There are blurred boundaries between lactic acidosis *per se* and diabetic ketoacidosis with raised lactate levels in that all cases of phenformin lactic acidosis also have raised levels of ketone bodies and indeed the ketoacids may account for as much as 50% of the total organic acid load. However, it is important to distinguish those cases where uncontrolled diabetic ketoacidosis is the primary problem in that lactate levels fall sharply on instituting fluid and small-dose insulin therapy and large amounts of bicarbonate are not required (and may indeed be contraindicated). The reasons for grossly raised blood lactate concentrations in certain cases of decompensated diabetes are not definitely known, but it seems probable that they are secondary either to circulatory problems or to the impaired ability of the acidotic liver to metabolize lactate.

We have also found that blood lactate concentrations rise in man and animals during the initial hours of therapy if large doses of insulin are used [24]. This effect is not found with low dose therapy, and is presumably due to insulin-induced increased peripheral lactate production at the same time as gluconeogenesis is inhibited by insulin.

Bicarbonate therapy may itself have profound effects on lactate metabolism. In experimental ketoacidosis, we have found that giving bicarbonate alone leads to a highly significant decrease in hepatic and muscle ATP with clear evidence of tissue anoxia [14]. This is associated with a sharp rise in liver and blood lactate concentrations. This is probably secondary to changes in oxygen delivery to the tissues. In severe acidosis, the lowered pH shifts the oxyhaemoglobin dissociation curve to the right while lowered red cell 2,3-diphosphyglycerate concentrations shift the curve to the left. The net effect is no change in the P_{50}. When, however, pH is corrected rapidly with bicarbonate, the correction of 2,3-diphosphoglycerate lags behind and the dissociation curve is shifted to the left with a marked fall in P_{50}, i.e. there is potentially impaired oxygen delivery. In a clinical sense, this implies that bicarbonate should be used with caution. It would be of interest to monitor lactate and lactate/pyruvate ratios in such situations.

B. Stable Diabetes

Fasting blood lactate in insulin-dependent diabetics is often within the normal range or only moderately elevated. A dramatic change occurs,

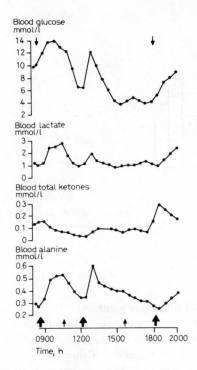

Fig. 6. Metabolic rhythm in a 15-year-old insulin-dependent diabetic treated by Actrapid and Montard insulins twice daily. The arrows at the top indicate insulin administration. Other symbols as in figure 2.

however, following breakfast and subcutaneous insulin. ALBERTI *et al.* [1] showed a mean increase in blood lactate concentrations of 100% following breakfast and insulin which they ascribed to the effect of pharmacological rather than physiological concentrations of insulin affecting glycolysis [2]. Many patients, however, show an increase in blood lactate concentration far in excess of this. We have seen patients in whom the concentration raises from a normal fasting value to 8 mmol/l following the morning insulin and breakfast [3a] although a rise of this magnitude is exceptional.

Figure 6 shows a rather more typical rise. Despite attempts to achieve good control of this 15-year-old girl's diabetes, blood glucose and total blood ketone bodies concentration were clearly abnormal in the fasting state. Blood lactate concentration at this time was normal, however, and it was only after her morning injection of Actrapid and Monotard insulin with breakfast that an abnormal rise in lactate concentration occurred. No

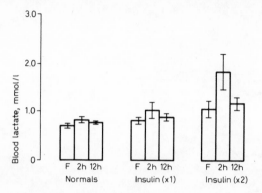

Fig. 7. Blood lactate concentration (mean±SEM) in normal subjects and diabetics treated by insulin once daily (×1) or twice daily (×2). Symbols as in figure 3.

rise is seen in blood lactate after lunch although a similar pattern to the morning was seen after her evening injection of insulin. Although part of the rise in lactate may be due to glucose utilization, it is probable that it is also due in part to inhibition of lactate uptake for hepatic glucose production. As can be seen in figure 7, this pattern of response is by no means atypical. In another patient, we have consistently found raised blood lactate concentrations in the face of normal blood glucose levels. It is not clear whether insulin administration to such patients appropriate to their blood glucose concentration restores blood lactate concentration to normal. We have sought an answer to this question by using a glucose-controlled insulin infusion system (GCIIS – Biostator system). Using a GCIIS, intravenous insulin is administered according to the patients' blood glucose concentration in order to maintain it in the normal range. Results from our preliminary studies [9] suggest that when glucose is maintained within the normal range, blood lactate concentration is also within the normal range. It should be stressed, however, that the maintenance of blood glucose within the normal range does not necessarily result in a normalization of the diurnal pattern for blood glucose. In addition, the GCIIS administers insulin into the peripheral blood system rather than the portal system. We have indeed found that blood lactate concentrations are returned towards normal when insulin is given sub-cutaneously as a continuous infusion using a preprogrammed pump [36]. These results emphasize the importance of careful planning of insulin therapy for all diabetics if normality in a broad metabolic sense is desired.

Lactic Acidosis in Diabetics

Cohen and Woods [10] collated reports on 104 patients who developed lactic acidosis. From these patients, type A acidosis was present in 39, and of the 65 remaining cases 39 also had diabetes. Of these 39, 28 were taking phenformin, leaving only 11 cases reported with the association between lactic acidosis and diabetes mellitus.

At present, there is no clear indication that diabetes mellitus in the absence of phenformin carries an increased risk of developing type B lactic acidosis, although diabetics do have an increased chance of having conditions, particularly vascular disease, which may predispose to the development of type A lactic acidosis.

Lactic Acidosis during Biguanide Therapy

At the time of writing, there are at least 372 cases of lactic acidosis during biguanide therapy reported in the literature [7, 30]. Of these, phenformin has been responsible for the majority of cases and metformin for only 18 cases. That this difference is not related solely to the amounts prescribed has been stressed by Berger and Amrein [7] who have pointed to the low incidence of biguanide-induced lactic acidosis in countries where metformin is the most heavily prescribed biguanide.

This is perhaps not surprising in view of tissue handling and metabolism of the biguanide. Phenformin is concentrated by the liver and inactivated by hydroxylation in a microsomal enzyme system. In contrast, both metformin and buformin are not metabolized to any appreciable extent but are excreted by the kidney [5, 6]. Buformin, however, has been ascribed a causal role in 64 cases of lactic acidosis which is a number far in excess of metformin. It would seem that buformin shares the disadvantages of phenformin despite similar tissue handling to metformin. In addition, a study of the presentation of buformin lactic acidosis shows no differences from phenformin lactic acidosis which is in contradistinction to metformin-induced cases where serum creatinine is significantly raised (table I). Indeed, of the reported cases of lactic acidosis during metformin therapy most of these could have been avoided with minimal consideration of the mechanisms of action of the drug. Thus, the administration of metformin to anuric patients is difficult to defend, a fact which the authors readily admit [4].

Our own diurnal studies support the concept of an essential difference between phenformin and metformin. Thus, for doses of the drugs which produced similar blood glucose concentrations, phenformin therapy re-

Table I. Laboratory findings at diagnosis (mean ± SEM) of biguanide-induced lactic acidosis and mortality. Data compiled from LUFT et al. [30] and BERGER and AMREIN [7]. Figures in parentheses are the number of patients on whom data was available

Biguanide	Number of patients	% mortality	pH	Blood glucose mg/100 ml	Blood lactate mmol/l	Serum creatinine mg/100 ml	Blood urea mg/100 ml
Phenformin	291	52.8	6.95 ± 0.11 (246)	176 ± 9 (219)	17.3 ± 0.6 (231)	2.9 ± 0.2 (88)	114 ± 4 (174)
Buformin	64	45.0	6.97 ± 0.15 (60)	176 ± 16 (54)	15.9 ± 1.0 (45)	3.9 ± 0.3 (52)	126 ± 9 (36)
Metformin	18	15.4	7.04 ± 0.43 (8)	166 ± 42 (13)	12.5 ± 1.3 (12)	7.6 ± 1.6 (6)	112 ± 21 (4)

sulted in higher concentrations of lactate than metformin [32]. If, as we stated above, the hyperlactataemic effect of the biguanide are inseparable from their hypoglycaemic effect, it is not easy to see how this difference can occur. At the subcellular level, however, SCHÄFER [39] has shown that the effects upon the mitochondrial membrane are related to the length of the side chain on the biguanide moiety with the longer side chain of phenylethylbiguanide having a greater effect than dimethylbiguanide.

To a certain extent, the difference has become of academic interest only with the withdrawal of phenformin from the market in several countries. In general terms, we agree with the tendency to use metformin as the biguanide of choice although with certain reservations. It is still necessary to consider carefully whether contraindications to its use are exhibited by the patient. More importantly, metformin still produces a chronic, moderate elevation in a number of intermediary metabolites, particularly the gluconeogenic precursors lactate, pyruvate and alanine. Although the significance of this is uncertain, it is clear that the use of metformin, while correcting one abnormality (blood glucose concentration), produces by its mode of action several other metabolic abnormalities.

Treatment of Lactic Acidosis

The treatment of lactic acidosis is similar whatever its aetiology. It is based on identification of the precipitating factor and treatment of this and the acidosis. With biguanide-induced lactic acidosis, the treatment is directed towards supporting the patient until the drug can be eliminated.

Bicarbonate is given in large amounts to correct the acidosis. Its clinical efficacy is by no means confirmed and in the retrospective analysis of LUFT et al. [30] there was no difference between the amounts of bicarbonate given to survivors and non-survivors although COHEN and WOODS [10] were able to detect a trend towards a lower mortality with the restoration of pH to normal by bicarbonate. As both authors point out, however, much of the reporting is anecdotal and subject to human frailty. There can be little doubt that a successful outcome is more likely to be reported than an unsuccessful one. There are strong theoretical grounds, however, for attempting restoration of pH to near-normal. LLOYD et al. [29] have shown the change that occurs in the liver with developing acidosis. Thus, at a pH below 7.1 the liver ceases to be the major site of lactate utilization and becomes a lactate-producing organ [29]. In the face of this, the majority of physicians would attempt to raise the pH above 7.1. The administration of bicarbonate in such quantities results often in sodium overload which may necessitate dialysis against a low-sodium dialysate. Although a number of peritoneal dialysis solutions contain lactate, the ion is not itself responsible for the acidosis and such solutions are not contraindicated although alternatives such as acetate are perhaps preferable. Haemodialysis against bicarbonate solutions is a useful way of alkalinizing the patient, although the concomitant removal of lactate ion is of no particular value. It may be a means of removing the biguanide, although it has been tried in too few cases to reach firm conclusions [4, 11, 18, 22, 40].

Insulin and dextrose infusion is based upon the theoretical concept of stimulating pyruvate dehydrogenase activity. While it is certainly necessary to correct the hypoglycaemia with which the patient may present, it is clear from animal experiments that it may worsen the acidosis by stimulating glycolysis [26].

The hydrogen acceptor methylene blue and the amine buffer trishydroxymethylaminomethane have both been used to a limited extent to treat lactic acidosis. They are probably of historical interest and the reader is referred to the discussion by COHEN and WOODS [10].

Dichloroacetate has theoretical advantages in the management of lactic acidosis. Its action is to stimulate pyruvate dehydrogenase activity and hence lower circulating concentration of pyruvate and lactate [44]. This lactate-lowering effect has been observed in diabetic man [41], and in animals with diabetic ketoacidosis [8] and mild lactic acidosis [25]. Figure 8 shows the results obtained from infusion of DCA 100 mg/kg into a normal subject. It is too early to say whether dichloroacetate therapy has significant benefits in lactic acidosis, but a clinical trial would appear warranted.

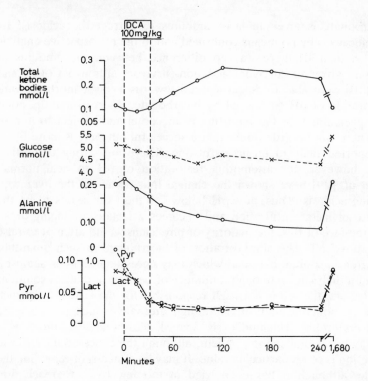

Fig. 8. Effect of intravenous dichloroacetate (DCA) infusion (100 mg/kg) upon blood metabolite concentrations in a normal subject.

Conclusions

Insulin has profound effects on carbohydrate metabolism and it is to be expected that lactate metabolism would be deranged in insulinopenic diabetics. Surprisingly, there is little evidence for abnormal blood lactate concentrations in such diabetics, most of the abnormalities in diabetics being iatrogenic.

Biguanides have a clear hyperlactataemic effect in all subjects and this may lead to the development of lactic acidosis in some subjects. Present methods of insulin therapy also induce grossly abnormal blood lactate patterns in many subjects, which can be eliminated by use of glucose-insulin feedback systems. Lactate concentrations may be raised in diabetic ketoacidosis but respond well to rehydration and low dose insulin therapy. Vigorous use of bicarbonate may, however, further elevate values due to interference with oxygen delivery to tissues, and should be used with caution.

Acknowledgements

We would like to thank DR. P. G. TODD for allowing us to study his patients, and Life Science Instruments; Miles Laboratories; Elkart, Indiana for the loan of the Biostator system. Financial support from the British Diabetic Association and Hoechst UK Ltd. is gratefully acknowledged.

References

1   ALBERTI, K. G. M. M.; DORNHORST, A., and ROWE, A. S.: Metabolic rhythms in old age. Biochem. Soc. Trans. *3:* 132–133 (1975).
2   ALBERTI, K. G. M. M.; DORNHORST, A., and ROWE, A. S.: Metabolic rhythms in normal and diabetic man. Israel J. med. Scis *11:* 571–580 (1975).
3a ALBERTI, K. G. M. M. and NATTRASS, M.: Metabolic abnormalities in juvenile diabetics; in AAGENAES Diabetes og diabetes-behandling II, pp. 125–137 (Lindgren & Söner, Mölndal 1977).
3b ALBERTI, K. G. M. M. and NATTRASS, M.: Metabolic disorders of diabetes and their management; in BESSER Advanced Medicine, vol. 13, pp. 173–189 (Pitman, Tunbridge Wells 1977).
4   ASSAN, R.; HEUCLIN, C.; GANEVAL, D.; BISMUTH, C.; GEORGE, J., and GIRARD, J. R.: Metformin-induced lactic acidosis in the presence of acute renal failure. Diabetologia *13:* 211–217 (1977).
5   BECKMAN, R.: Über die Resorption und den biologischen Abbau von 1-(β-Phenäthyl)-biguanid (Phenformin). Diabetologia *3:* 368–376 (1966).
6   BECKMAN, R.: Resorption, Verteilung im Organismus und Ausscheidung von Metformin. Diabetologia *5:* 318–324 (1969).
7   BERGER, W. und AMREIN, K.: Laktatazidosen unter der Behandlung mit den 3 Biguanidpräparaten Phenformin, Buformin und Metformin. Schweizerische Rundschau für Medizin 'Praxis' (1978).
8   BLACKSHEAR, P. J.; HOLLOWAY, P.A.H., and ALBERTI, K. G. M. M.: Metabolic interactions of dichloroacetate and insulin in experimental diabetic ketoacidosis. Biochem. J. *146:* 447–556 (1975).
9   BUCKLE, A. L. J.; NATTRASS, M.; CLUETT, B.; STUBBS, W. A.; WALTON, R. J.; ALBERTI, K. G. M. M., and CLEMENS, A. H.: Blood metabolite concentrations in diabetics: effect of normalisation of blood glucose using a glucose controlled insulin infusion system (GCIIS). Diabetologia *13:* 385 (1977).
10 COHEN, R. D. and WOODS, H. F.: Clinical and biochemical aspects of lactic acidosis (Blackwell, Oxford 1976).
11 COHEN, R. D.; WARD, J. D.; BRAIN, A. J. S.; MURRAY, C. R.; SAVEGE, T. M., and ILES, R. A.: The relation between phenformin therapy and lactic acidosis. Diabetologia *9:* 43–46 (1973).
12 COORE, H. G.; DENTON, R. M.; MARTIN, B. R., and RANDLE, P. J.: Regulation of adipose tissue pyruvate dehydrogenase by insulin and other hormones. Biochem. J. *125:* 115–127 (1971).
13 CRAIG, J. W.; MILLER, M.; WOODWARD, H., and MERIK, E.: Influence of phenethylbiguanide on lactic, pyruvic and citric acids in diabetic patients. Diabetes *9:* 186–193 (1960).
14 CUTHBERT, C. and ALBERTI, K. G. M. M.: Acidaemia and insulin resistance in the diabetic ketoacidotic rat. Metabolism (in press, 1978).

15 DENTON, R. M.; RANDLE, P. J.; BRIDGES, B. J.; COOPER, R. H.; KERBEY, A. L.; PASK, H. T.; SEVERSON, D. L.; STANSBIE, D., and WHITEHOUSE, S.: Regulation of mammalian pyruvate dehydrogenase. Molec. cell. Biochem. 9: 27–53 (1975).

16 DIETZE, G.; WICKLMAYR, M.; HEPP, K. D.; BOGNER, W.; MEHNERT, H.; CZEMPIEL, H., and HENFTLING, H. G.: On gluconeogenesis of human liver. Accelerated hepatic glucose formation induced by increased precursor supply. Diabetologia 12: 555–561 (1976).

17 DOAR, J. W. H. and CRAMP, D. G.: The effects of obesity and maturity-onset diabetes mellitus on L(+) lactic acid metabolism. Clin. Sci. 39: 271–279 (1970).

18 EWY, G. A.; PABICO, R. C.; MAHER, J. F., and MINTZ, D. H.: Lactate acidosis associated with phenformin therapy and localized tissue hypoxia. Ann. intern. Med. 59: 878–883 (1963).

19 FAJANS, S. S.; MOORHOUSE, G. A.; DOORENBOS, H.; LOUIS, L. H., and CONN, J. W.: Metabolic effects of phenethylbiguanide in normal subjects and in diabetic patients. Diabetes 9: 194–201 (1960).

20 FELIG, P.: The glucose-alanine cycle. Metabolism 22: 179–207 (1973).

21 GRIES, F. A.: Hormone control of human adipose tissue metabolism in vitro; in JEANRENAUD and HEPP Adipose tissue regulation and metabolic functions, pp. 161–171 (Academic Press, New York 1970).

22 HAYAT, J. C.: The treatment of lactic acidosis in the diabetic patient by peritoneal dialysis using sodium acetate. A report of two cases. Diabetologia 10: 485–487 (1974).

23 HERS, H. G.: The control of glycogen metabolism in the liver. Annu. Rev. Biochem. 45: 167–189 (1976).

24 HOCKADAY, T. D. R. and ALBERTI, K. G. M. M.: Diabetic coma. Clin. Endocr. Metab. 1: 751–788 (1972).

25 HOLLOWAY, P. A. H. and ALBERTI, K. G. M. M.: Dichloroacetate – a therapy for phenformin-induced lactic acidosis. Diabetologia 13: 402 (1977).

26 IVERSEN, J.; HOLLOWAY, P. A. H., and ALBERTI, K. G. M. M.: Phenformin-induced lactic acidosis: ineffectiveness of insulin and glucose therapy. Eur. J. clin. Invest. 6: 317 (1976).

27 KERBEY, A. L.; RANDLE, P. J.; COOPER, R. H.; WHITEHOUSE, S.; PASK, H. T., and DENTON, R. M.: Regulation of pyruvate dehydrogenase in rat heart. Biochem. J. 154: 327–348 (1976).

28 LEVINE, R.; GOLDSTEIN, M. S.; HUDDLESTON, B., and KLEIN, S. P.: Action of insulin on the permeability of cells to free hexoses, as studied by its effects on the distribution of galactose. Am. J. Physiol. 163: 70–76 (1950).

29 LLOYD, M. H.; ILES, R. A.; SIMPSON, B. R.; STRUNIN, J. M.; LAYTON, J. M., and COHEN, R. D.: The effect of simulated metabolic acidosis on intracellular pH and lactate metabolism in the isolated perfused rat liver. Clin. Sci. mol. Med. 45: 543–549 (1973).

30 LUFT, D.; SCHMÜLLING, R. M., and EGGSTEIN, M.: Lactic acidosis in biguanide-treated diabetics. Diabetologia 14: 75–87 (1978).

31 MALLETTE, L. E.; EXTON, J. H., and PARK, C. A.: Effects of glucagon on amino acid transport and utilisation in the perfused rat liver. J. biol. Chem. 244: 5724–5728 (1969).

32 NATTRASS, M.; TODD, P. G.; HINKS, L.; LLOYD, B., and ALBERTI, K. G. M. M.: Comparative effects of phenformin, metformin and glibenclamide on metabolic rhythms in maturity-onset diabetics. Diabetologia 13: 145–152 (1977).

33 NATTRASS, M.; TODD, P. G.; TURNELL, D., and ALBERTI, K. G. M. M.: Metabolic

abnormalities during combined sulphonylurea and phenformin therapy in maturity-onset diabetics. Diabetologia *14:* 389–395 (1978).

34   NATTRASS, M.; HINKS, L.; SMYTHE, P.; TODD, P. G., and ALBERTI, K. G. M. M.: Metabolic effects of combined sulphonylurea and metformin therapy in maturity onset diabetics. Diabète Métab. (in press).

35   PETTIT, F. H.; PELLEY, J. W., and REED, L. J.: Regulation of pyruvate dehydrogenase kinase and phosphatase by acetyl-CoA/CoA and NADH/NAD ratios. Biochem. biophys. Res. Commun. *65:* 575–582 (1975).

36   PICKUP, J.; KEEN, H.; PARSONS, J. A., and ALBERTI, K. G. M. M.: Continuous subcutaneous insulin infusion: an approach to obtaining normoglycaemia. Br. med. J. *i:* 204–207 (1978).

37   RANDLE, P. J.; GARLAND, P. B.; HALES, C. N., and NEWSHOLME, E. A.: The glucose and fatty acid cycle. Its role in insulin sensitivity and the metabolic disturbance of diabetes mellitus. Lancet *i:* 785–789 (1963).

38   SCHADE, D. S. and EATON, R. P.: Dose response to insulin in man: differential effects on glucose and ketone body regulation. J. clin. Endocr. Metab. *44:* 1038–1053 (1977).

39   SCHÄFER, G.: On the mechanism of action of hypoglycaemia-producing biguanides. A re-evaluation and a molecular theory. Biochem. Pharmacol. *25:* 2005–2014 (1976).

40   SELROOS, O.; PASTERNACK, A., and KUHLBÄCK, B.: Laktatacidos och fenformin. Nord. med. *80:* 1658–1661 (1968).

41   STACPOOLE, P. W.; MOORE, G. W., and KORNHAUSER, D. M.: Metabolic effects of dichloroacetate in patients with diabetes mellitus and hyperlipoproteinemia. New Engl. J. Med. *298:* 526–530 (1978).

42   WATKINS, P. J.; SMITH, J. S.; FITZGERALD, M. G., and MALINS, J. M.: Lactic acidosis in diabetes. Bri. med. J. *i:* 744–747 (1969).

43   WEBER, G.: Hormonal control of gluconeogenesis; in BITTAR and BITTAR The biological basis of medicine, vol. 2, pp. 263–307 (Academic Press, New York 1968).

44   WHITEHOUSE, S.; COOPER, R. H., and RANDLE, P. J.: Mechanism of activation of pyruvate dehydrogenase by dichloroacetate and other halogenated carboxylic acids. Biochem. J. *141:* 761–774 (1974).

M. NATTRASS, MD, Faculty of Medicine, Chemical Pathology and Human Metabolism, Level D. South Laboratory and Pathology Block, General Hospital, *Southampton* (England)
Present address: Department of Medicine, University of Chicago, 950 East 59th Street, *Chicago, IL 60637* (USA)

Lactate in Acute Conditions. Int. Symp., Basel 1978, pp. 102–114 (Karger, Basel 1979)

The Role of Liver Dysfunction in the Genesis of Lactic Acidosis[1]

H. F. Woods and H. Connor

Department of Pharmacology and Therapeutics, The University of Sheffield, Sheffield

Introduction

Cohen and Woods [1976] have made the distinction between those cases of lactic acidosis in which there is evidence of hypotension, poor tissue perfusion or arterial oxygen desaturation (type A) and those cases where these features are not present at the time of diagnosis (type B). Type B can be further subdivided (table I). Of the common disorders associated with type B lactic acidosis, diabetes and biguanide therapy are discussed by Alberti and Nattrass elsewhere in this book. This paper is concerned with the occurrence of lactic acidosis in patients with liver disease and with the role of the liver in the genesis of lactic acidosis.

Evidence of liver disease is found in 30–40% of patients with lactic acidosis [Tranquada et al., 1966; Cohen and Woods, 1976]. This association is explicable because the liver is known to be a major site of lactate utilization. Kramer et al. [1971] reported that, in anaesthetized dogs, 30% of an intravenous load of sodium L_+ lactate was removed by the liver. From the catheter studies of Rowell et al. [1966], it can be calculated that lactate uptake by the human liver is approximately 0.72 mol/24 h/70 kg BW. This represents between 30 and 50% of the normal daily production which has been variously estimated as 1.27 [Cohen and Woods, 1976], 1.53 [Kreisberg et al., 1970] and 1.95 [Searle and Cavalieri, 1972] mol/24 h/70 kg BW. The liver is therefore a major site of lactate utilization but can, in certain conditions, become an organ of lactate production. Examples of such conditions are impaired liver blood flow, liver hypoxia, hyperventilation [Berry and Scheuer,

---

[1] The studies of lactate utilisation in man were performed by H. C. during the tenure of an MRC Clinical Research Fellowship.

Table I. Classification of lactic acidosis. After COHEN and WOODS [1976]

Type A
Associated with arterial hypoxia and/or hypotension

Type B
Associated with
1. Common disorders
   Liver disease
   Renal disease
   Diabetes mellitus
   Leukaemias and reticuloses
2. Drugs and toxins
   Biguanides
   Sorbitol, fructose, xylitol
   Ethanol and methanol
3. Hereditary types

1967] and fructose infusion [WOODS et al., 1970; BERGSTRÖM et al., 1973]. BERRY [1967] emphasized the potential importance in the genesis of lactic acidosis of an organ which could change from being a major site of lactate utilization to a site of lactate production and he concluded that: 'In the present state of our knowledge it would seem advisable to regard every patient suffering from lactic acidosis as a potential case of hepatic failure'.

In this paper, BERRY's conclusion will be re-examined in the light of subsequent investigations.

Utilization of Sodium L_+ Lactate in Patients with Liver Disease

If the liver is indeed a major site of lactate utilization, it is to be expected that the rate of removal of an intravenous load of sodium L_+ lactate will be impaired in patients with hepatic disease. However, the evidence concerning this point has been conflicting. In patients with either infective hepatitis or arsphenamine jaundice, SOFFER et al. [1938] reported persistent elevation of lactate concentrations in the blood 30 min after a bolus injection. However, RECORD et al. [1975a] reported that the clearance of lactate from blood was normal in patients with paracetamol-induced hepatitis and that the hyperlactataemia in these patients was due to increased lactate production, not to decreased removal. Further studies have recently been reported by CONNOR et al. [1978], who infused molar sodium L_+ lactate at a rate of 0.075 mM/kg BW/min for 20 min into normal subjects and a group of patients with liver disease. Mathematical

Table II. The kinetics of lactate metabolism in normal subjects and patients with liver disease (mean±SEM) [CONNOR et al., 1978]

	Pre-infusion blood lactate concentration mmol/l	Clearance ml/min/kg BW	Endogenous production rate of lactate mol/24 h/70 kg BW
Normal (n = 13)	0.77±0.10	18±2	1.59±0.29
Liver disease (n = 35)	1.12±0.07	14±1	1.58±0.13
p	<0.01	<0.01	n.s.

models were used to calculate the elimination constant (K) and the apparent volume of distribution (V) of the infused lactate in each subject. From this information, it was possible to calculate the blood clearance of lactate and the endogenous production rate of lactate at rest. The results are summarized in table II, from which it can be concluded that the hyperlactataemia in patients with liver disease was a consequence of decreased lactate utilization; endogenous production was not significantly higher than that in the normal subjects. The type of liver disease in these patients included chronic active hepatitis, alcoholic liver disease, primary biliary cirrhosis, intrahepatic cholestasis, biliary obstruction and secondary malignancy.

Abnormalities of lactate utilization were found in all subgroups except those with cholestatic hepatitis; patients with acute viral hepatitis were not studied. Values for elimination $T\frac{1}{2}$ of up to three times the normal mean were found in some patients with alcoholic liver disease and chronic active hepatitis.

The most likely cause of the impaired lactate utilization in liver disease is reduction in the size of the functional hepatocyte mass but it is also possible that the presence of liver disease might alter lactate utilization in other tissues. This possibility cannot be refuted on the basis of existing evidence from human studies and would be analogous to the finding of impaired glucose tolerance in liver disease. However, HENLEY et al. [1973, 1974] have shown impaired gluconeogenesis from lactate in the perfused liver of rats which have been fed on a diet designed to induce hepatic steatosis or cirrhosis. This finding would support the conclusion that diminished lactate utilization in human hepatic disease is a result of decreased lactate metabolism in the liver itself.

The Pathogenesis of Lactic Acidosis

Hyperlactataemia is a common finding in patients with liver disease [ALBERTI, 1975], but only a small proportion of such patients will develop lactic acidosis; indeed, alkalosis is a more common finding than acidosis [RECORD et al., 1975b; MULHAUSEN et al., 1967]. The hyperlactataemia can be explained by the impaired lactate utilization described in the preceding section, but it is clear that lactic acidosis is not an invariable consequence of diminished lactate removal in patients with liver disease. Before lactic acidosis can develop, it appears that the defect in hepatic metabolism must be very severe, or other metabolic defects must be present. Examples of other possible abnormalities include: (a) increased lactate production; (b) defective removal of hydrogen ion; (c) impaired lactate utilization in tissues other than liver. These possibilities will be discussed in turn.

Increased Lactate Production

An increase in lactate production is probably the main contributing factor to the lactic acidosis associated with shock and hypoxia (type A lactic acidosis), but impaired lactate utilization is also involved. ELDRIDGE et al. [1974] measured lactate turnover with $[1-^{14}C]$ lactate in dogs and infused unlabelled sodium L_+ lactate so as to determine lactate turnover at different blood lactate concentrations. Lactate turnover has also been estimated before and during the induction of haemorrhagic shock [EL-DRIDGE, 1974]. Lactate turnover at a given blood lactate concentration in the shocked animals was lower than would be expected from the experiments in which sodium L_+ lactate was infused. This suggests that impaired lactate utilization is involved in lactic acidosis associated with haemorrhagic shock. Similar findings in hypovolaemic shock and in experimental cardiac tamponade in dogs have been reported by SCHRÖDER et al. [1969] and by DANIEL et al. [1974]. Not only is there this defect in lactate removal but the organs which usually make the major contribution to lactate removal may actually produce lactate when shock is present. Thus, BERRY and SCHEUER [1967] reported that the dog liver invariably produced lactate when a mean systolic pressure of 147 mm Hg was lowered to a mean value of 43 mm Hg by venesection. SRIUSSADAPORN and COHN [1968] found renal production of lactate in dogs in hypovolaemic shock and in 4 of 8 patients with type A lactic acidosis. It is clear then that both overproduction and under-utilization can contribute to the lactic acidosis of haemorrhagic shock and that part of the excess production can originate in organs which normally metabolize lactate.

Infusions of fructose, sorbitol or xylitol have all been associated with

Table III. The kinetics of lactate metabolism in normal subjects and patients with secondary malignancy of the liver (mean±SEM)

	Pre-infusion blood lactate concentration mmol/l	Clearance ml/min/kg BW	Endogenous production rate of lactate mol/24 h/70 kg BW
Normal (n = 13)	0.77±0.10	18±2	1.59±0.29
Liver malignancy (n = 4)	1.72±0.27	13±1	2.20±0.34
p	<0.005	<0.05	<0.05

lactic acidosis in man [COHEN and WOODS, 1976]. Measurement of arteriovenous lactate concentrations across the liver have not been made in these conditions, but it can be predicted from a knowledge of fructose metabolism that the infusion of fructose or sorbitol will cause production of lactate in the liver [WOODS and ALBERTI, 1972], and an increase in lactate production has been shown in the isolated perfused rat liver when xylitol was present in the perfusion medium [WOODS and KREBS, 1973]. In some cases where lactic acidosis develops during infusion of these substrates, there may be existing liver and kidney disease [CONNOR, 1978] so that, as in hypovolaemic shock, a combination of under-utilization and over-production of lactate is probable. This same combination has been demonstrated in patients with metastatic carcinoma of the liver (table III).

Impaired Hydrogen Ion Metabolism

The lactic acid which is produced in the body is almost completely ionized because its pK (3.8) is well below physiological pH. The metabolism of the lactate ion, whether by oxidation or gluconeogenesis, also involves the metabolism of hydrogen ion:

$$2CH_3CHOHCOO^- + 2H^+ + 6O_2 \rightarrow 6CO_2 + 6H_2O,$$
$$2CH_3CHOHCOO^- + 2H^+ \rightarrow C_6H_{12}O_6.$$

Thus, hyperlactataemia, whether due to increased lactate production or decreased lactate removal, should imply an equivalent excess of hydrogen ions. It is evident, however, that some of the hydrogen ions can either be buffered or can be removed by urinary excretion, thus explaining the finding of hyperlactataemia despite normal or even alkalotic pH in many patients with liver disease [RECORD et al., 1975b; MULHAUSEN et al., 1967]. A defect in buffering capacity or in renal excretion of acid could therefore contribute to the onset of lactic acidosis.

Two patients who developed lactic acidosis at the time of infusions with a sorbitol-ethanol-amino acid mixture had each had a previous unexplained episode of metabolic acidosis before the infusions were started [CONNOR, 1978]. This suggests that there was a pre-existing abnormality in hydrogen ion metabolism in these patients before the onset of lactic acidosis. An impaired ability to excrete an acid load has been described during phenformin therapy and may contribute to the genesis of lactic acidosis in some patients on this drug [ROOTH and BANDMAN, 1973].

Impaired Lactate Utilization in Extrahepatic Tissues
1. Resting Skeletal Muscle
KREISBERG [1974] has suggested that failure of lactate uptake by the liver will not result in lactic acidosis because uptake by resting skeletal muscle will compensate for the hepatic defect. Skeletal muscle normally produces lactate but inherent in KREISBERG'S suggestion is the implication that resting skeletal muscle will switch to lactate utilization if the blood lactate concentration is elevated. However, evidence for lactate utilization in resting muscle does not exist. AHLBORG et al. [1976] have demonstrated lactate uptake by 'resting' limbs during infusion of sodium lactate, but the sodium load in these experiments was sufficient to increase limb blood flow by 160%. The muscle was not therefore in its physiologically resting state even though the subjects studied were not exercising. The same limitation applies to reports of positive arteriovenous differences in lactate concentrations in non-exercising limbs during short periods of exercise [HARRIS et al., 1962; FREYSCHUSS and STRANDELL, 1967], because there is some increase in blood flow in the non-exercising limb. Moreover, in none of these reports is any evidence presented to indicate whether resting muscle is actually metabolizing lactate or whether it is merely acting as a temporary storage site. The evidence of JACKSON et al. [1972] would suggest the latter. Even if resting skeletal muscle can make a contribution to lactate removal, it is known from the studies in patients with hepatic disease described in this paper that skeletal muscle cannot compensate for impaired hepatic uptake of lactate. Admittedly, the conditions during infusion of sodium lactate are different from those found in lactic acidosis; in particular the infusion of sodium lactate results in alkalosis. The effects of changes in pH on lactate removal by different organs are discussed below.

Table IV. The kinetics of elimination of an L_+ lactic acid load in rats subjected to either bilateral nephrectomy or sham operation (mean ± SEM). Data from YUDKIN and COHEN [1975]

	Elimination half-life min	Blood clearance ml/min/kg BW
Sham operation (n = 9)	5.28 ± 0.21	45.3 ± 4.5
Bilateral nephrectomy (n = 10)	7.10 ± 0.48	32.0 ± 2.8
p	< 0.01	< 0.05

Table V. Kinetics of lactate metabolism in anephric man (mean ± SEM)

	Pre-infusion blood lactate concentration mmol/l	Elimination half-life min	Clearance ml/min/kg BW
Anephric			
1	0.65	10.7	18.3
2	0.66	11.5	15.5
3	0.36	13.5	18.6
Average	0.56	11.9	17.8
Normal range	0.4–1.2	8.5–16	12–30

2. Kidneys

Lactate can be metabolized by the perfused kidney [NISHIITSUTSUJI-UWO *et al.*, 1967] and removal of an intravenous load of lactic acid is slower in rats subjected to bilateral nephrectomy than in animals who have had a sham operation (table IV) [YUDKIN and COHEN, 1975]. This indicates that the kidneys can make a significant contribution to the removal of an L_+ lactic acid load and that removal of lactate by skeletal muscle, if it occurs, cannot compensate for the loss of the kidneys.

Removal of a load of sodium L_+ lactate was examined in 3 patients on maintenance haemodialysis in whom the kidneys had been removed. Lactate utilization in these subjects was normal (table V). Although this might imply a species difference between rat and man it is more likely to reflect a difference in blood pH in the two studies. The rats became acidotic during infusion of L_+ lactic acid whereas the humans became alkalotic during infusion of sodium L_+ lactate. As will be discussed, acidosis is associated with an increased renal contribution to lactate removal.

The Effect of pH on Lactate Removal by Liver and Kidneys. The studies of COHEN and his colleagues [LLOYD *et al.*, 1973; YUDKIN and COHEN, 1975] have shown how the relative contribution of different organs to lactate removal can be markedly altered by changes in pH. Lactate uptake by the isolated perfused rat liver falls when perfusate pH is less than 7.1–7.2, and if perfusate pH is lowered to below 7.0 there is net production of lactate by the liver [LLOYD *et al.*, 1973].

This contrasts with the effect of acidosis on lactate uptake by the kidney. YUDKIN and COHEN [1975] infused half-neutralized lactic acid into rats which had been subjected to either a bilateral nephrectomy or sham operation and then made acidotic with ammonium chloride. In these conditions, it was calculated that the contribution of the kidney to lactate removal rose from 16% at pH 7.45 to 44% at pH 6.75. The increase in renal lactate removal compensated for about half of the slowing in extrarenal lactate removal which occurred in acidosis.

It would appear that muscle and kidneys are unable to compensate for the presence of liver disease at normal or alkalaemic pH, and that muscle and liver are unable to compensate for the absence of kidneys at acidaemic pH. The effect of acidosis on removal of lactate by exercising skeletal muscle is not known, but myocardial lactate removal in anaesthetized dogs falls by 30% when pH is lowered to 7.16–7.20 by the infusion of hydrochloric acid [GOODYER *et al.*, 1961]. In patients with lactic acidosis, most of the skeletal muscle is not exercising but is at rest and, as has been discussed, there is no good evidence that it is removing lactate from blood even if the blood lactate concentration is elevated. There is no reason to believe that the onset of acidosis would induce resting muscle to take up lactate, although acidosis might lower lactate production in such muscle by inhibiting phosphofructokinase and thereby slowing glycolysis [UI, 1966].

Discussion

The Role of the Kidney

The evidence reviewed in this paper indicates that the liver and kidney are the major organs of lactate removal and that the relative contributions of these two organs depend on the pH. In both man and rat, it seems that the kidneys make little if any contribution to lactate removal at normal pH but become progressively more important as pH falls. It is therefore improbable that defective renal uptake of lactate is involved in the *initiation* of lactic acidosis unless the patient is already acidotic for some other reason. However, it is known that the majority of patients have evidence of abnormal glomerular function at the time when they develop lactic acidosis, so it is possible that there is some aspect of renal dysfunction (other than an impaired capacity for lactate uptake at acid pH) which contributes to the development of lactic acidosis. In phenformin-associated cases, this may be defective renal excretion of phenformin with consequent toxic drug levels. However, high blood urea concentrations are also common in patients with type B lactic acidosis who have not received phenformin; in 31 such cases where the blood urea concentration was recorded, it was greater than 7.5 mmol/l in 21 and greater than 16.7 mmol/l in 13 [COHEN and WOODS, 1976].

There are three other mechanisms by which renal disease could contribute to the initiation of lactic acidosis. One results from the impaired ability to excrete an acid load in cases of kidney damage; the effect of such a defect in a patient whose liver is already failing to metabolize lactate was discussed earlier. The two other possibilities result from the presence within the body of diseased kidneys and would not be mimicked by experiments involving nephrectomy. These possibilities are renal lactate production and an effect of diseased kidneys on lactate utilization in some remote organ such as the liver. Renal lactate production is well recognized as a contributing factor in type A lactic acidosis and its occurrence in shocked humans was documented by SRIUSSADAPORN and COHN [1968]. These same authors also found renal lactate production in a patient with lactic acidosis (pH 7.1, blood lactate 20 mmol/l) and normal blood pressure.

The possibility that the presence of diseased kidneys might alter lactate utilization in other organs does not seem to have been investigated. The kinetics of removal of a sodium lactate load have been studied in 2 patients with nephrotic syndrome [CONNOR et al., unpublished observations]. In one, the kinetics were completely normal but in the other the clearance was at the lower limit of normal (12.2 ml/kg BW/min) and elimination half-life was prolonged (17.5 min–normal 8.5–16 min). Abnormal lactate utilization might be yet another example of the many metabolic changes which occur in the nephrotic syndrome.

The Role of the Liver

The studies of BERRY and SCHEUER [1967] and ELDRIDGE [1974] in dogs show that impaired lactate removal by the liver contributes to the accumulation of lactate in shock, and SRIUSSADAPORN and COHN [1968] demonstrated net hepatic lactate production in 2 patients with type A lactic acidosis. Except for some congenital examples, there is no direct evidence for failure of hepatic lactate uptake in patients with type B lactic acidosis. Nevertheless, the demonstration of impaired utilization of a sodium lactate load in patients with liver disease taken in conjuction with the high incidence of liver disease in patients with lactic acidosis is strong indirect evidence. Moreover, the liver's contribution to lactate removal is greatest at normal or slightly acidotic pH which means that defective liver metabolism could be involved in the initiation of lactic acidosis. Once lactic acidosis has been initiated in this way, it is possible that a 'vicious circle' might develop because the fall in hepatic intracellular pH (as a consequence of impaired hepatic uptake of the lactate ion) will lower lactate uptake still further [COHEN et al., 1971]. If extracellular pH falls below 7.0, then the liver may switch to lactate production [LLOYD et al., 1973].

If it is assumed (table II) that a normal man of 70 kg BW has: (1) a blood lactate concentration of 0.77 mmol/l; (2) a lactate clearance of 1.26 l/min; (3) an endogenous production rate of 1.1 mmol/min, and a lactate space of 25.9 litres [CONNOR et al., unpublished observation], and if it is also assumed that the liver contributes between 30 and 50% to lactate removal and that no other organ can compensate for the hepatic contribution, then it can be calculated that the complete cessation of lactate uptake by the liver would result in a rise in blood lactate concentration of 0.7–1.2 mmol/l/h. However, this rise would not continue indefinitely. Within 1 h, a new steady-state lactate concentration of 1.2–1.7 mmol/l would be achieved. This concentration represents hyperlactataemia but not lactic acidosis which is usually associated with a blood lactate concentration of greater than 5 mmol/l. If a blood lactate concentration of 5 mmol/l were to result from defective hepatic removal of lactate alone, it would be necessary for liver lactate removal to constitute more than 80% of total lactate utilization. It is therefore unlikely that type B lactic acidosis could result from isolated failure of hepatic uptake of lactate. Indeed, a patient whose hepatic disease is severe enough to seriously lower lactate uptake might also be expected to produce lactate from damaged liver cells and to have disturbed renal metabolism. Type B lactic acidosis, like type A, is probably multifactorial in origin.

Conclusions

Evidence from both animals and man shows that lactic acidosis due to shock or hypoxia (type A) is the result of a combination of increased lactate production and decreased lactate removal. The theoretical considerations described in the preceding section suggest that type B lactic acidosis also results from a combination of factors, of which abnormal hepatic metabolism is one. The suggestion by BERRY [1967] that the liver occupies a central role in lactate homeostasis has been confirmed by further investigation. Thus, the liver is quantitatively the major site of lactate removal at normal pH. If liver disease is present, then other organs are unable to compensate for the hepatic contribution to lactate removal. Moreover, when acidosis is established it may further impair hepatic lactate uptake to the point that there is net lactate production by the liver. The kidney only makes a significant contribution to lactate removal when the patient is already acidaemic. It is therefore unlikely that defective renal utilization of lactate will contribute to the initiation of lactic acidosis unless there is a pre-existing acidosis. However, renal lactate consumption may help to limit the severity of lactic acidosis because the renal contribution to lactate removal increases as pH falls. Renal disease may contribute to the development of lactic acidosis in other ways; e.g. by defective excretion of hydrogen ion, by net lactate production (in both type A and type B cases) or possibly by a remote effect on lactate utilization in other organs.

References

AHLBORG, G.; HAGENFELDT, L., and WAHREN, J.: Influence of lactate infusion on glucose and FFA metabolism in man. Scand. J. clin. Lab. *36:* 193–201 (1976).

ALBERTI, K. G. G. M.: Some metabolic aspects of liver disease; in TRUELOVE and TROWELL Topics in gastroenterology, vol. 2, pp. 341–360 (Blackwell, Oxford 1975).

BERGSTRÖM, J.; FÜRST, P.; GALLYAS, F.; HULTMAN, E.; NILSSON, L. H.; ROCH-NORLUND, A. E., and VINNARS, E.: Aspects of fructose metabolism in normal man. Acta med. scand. *542:* suppl., pp. 57–64 (1973).

BERRY, M. N.: The liver and lactic acidosis. Proc. R. Soc. Med. *60:* 1260–1262 (1967).

BERRY, M. N. and SCHEUER, J.: Splanchnic lactic acid metabolism in hyperventilation, metabolic alkalosis and shock. Metabolism *16:* 537–547 (1967).

COHEN, R. D.; ILES, R. A.; BARNETT, D.; HOWELL, M. E. O., and STRUNIN, J.: The effect of changes in lactate uptake on the intracellular pH of the perfused rat liver. Clin. Sci. *41:* 159–170 (1971).

COHEN, R. D. and WOODS, H. F.: Clinical and biochemical aspects of lactic acidosis (Blackwell, Oxford 1976).

CONNOR, H.: The pathogenesis of lactic acidosis during parenteral nutrition; in BAXTER and JACKSON Clinical parenteral nutrition, (Geistlich Education Press, Chester 1978). pp. 226–241.

CONNOR, H.; WOODS, H. F.; MURRAY, J. D., and LEDINGHAM, J. G. G.: The kinetics of elimination of a sodium L-Lactate load in man: the effect of liver disease. Clin. Sci. mol. Med. *54:* 33–34P (1978).

DANIEL, A.; PIERCE, C. H.; SHIZGAL, H. M., and MACLEAN, L. D.: Lactate turnover in oligemic and normovolemic shock. Surg. Forum *25:* 3–4 (1974).

ELDRIDGE, F. L.: Relationship between lactate turnover rate and blood concentration in hemorrhagic shock. J. appl. Physiol. *37:* 321–323 (1974).

ELDRIDGE, F. L.: Relationship between lactate turnover rate and blood concentration in hemorrhagic shock. J. sppl. Physiol. *37:* 321–323 (1974).

FREYSCHUSS, U. and STRANDELL, T.: Limb circulation during arm and leg exercise in supine position. J. appl. Physiol. *23:* 163–170 (1967).

GOODYER, A. V. N.; ECKHARDT, W. F.; OSTBERG, R. H., and GOODKIND, M. J.: Effects of metabolic acidosis and alkalosis on coronary blood flow and myocardial metabolism in the intact dog. Am. J. Physiol. *200:* 628–632 (1961).

HARRIS, P.; BATEMAN, M., and GLOSTER, J.: The regional metabolism of lactate and pyruvate during exercise in patients with rheumatic heart disease. Clin. Sci. *23:* 545–560 (1962).

HENLEY, K. S.; LAUGHREY, E. G., and CLANCY, P. E.: Gluconeogenesis in the fatty liver of the rat: the effect of oleate and ethanol. J. Lab. clin. Med. *82:* 419–431 (1973).

HENLEY, K. S.; LAUGHREY, E. G., and CLANCY, P. E.: Gluconeogenesis in the cirrhotic liver of the rat. The effect of oleate or ethanol. J. Lab. clin. Med. *83:* 175–188 (1974).

JACKSON, R. A.; PETERS, N.; ADVANI, U.; PERRY, G.; ROGERS, J.; BROUGH, W. H., and PILKINGTON, T. R. E.: Forearm glucose uptake during the oral glucose tolerance test in normal subjects. Diabetes *22:* 442–458 (1973).

KRAMER, K.; DRIESSEN, G., and BRECHTELSBAUER, H.: Lactate elimination and O_2 consumption of the liver in narcotised dogs. Pflügers Arch. *330:* 195–205 (1971).

KREISBERG, R. A.: Mechanism of lactic acidosis. Lancet *ii:* 960 (1974).

KREISBERG, R. A.; PENNINGTON, L. F., and BOSHELL, B. R.: Lactate turnover and gluconeogenesis in normal and obese humans: effect of starvation. Diabetes *19:* 53–63 (1970).

LLOYD, M. H.; ILES, R. A.; SIMPSON, B. R.; STRUNIN, J. M.; LAYTON, J. M., and COHEN, R. D.: The effect of simulated metabolic acidosis on intracellular pH and lactate metabolism in the isolated perfused rat liver. Clin. Sci. mol. Med. *45:* 543–549 (1973).

MULHAUSEN, R.; EICHENHOLZ, A., and BLUMENTALS, A.: Acid-base disturbances in patients with cirrhosis of the liver. Medicine, Baltimore *46:* 185–189 (1967).

NISHIITSUTSUJI-UWO, J. M.; ROSS, B. D., and KREBS, H. A.: Metabolic activities of the isolated perfused rat kidney. Biochem. J. *103:* 852–862 (1967).

RECORD, C. O.; CHASE, R. A., and WILLIAMS, R.: Lactate metabolism in patients with paracetamol induced liver damage. Clin. Sci. mol. Med. *49:* 26P (1975a).

RECORD, C. O.; ILES, R. A.; COHEN, R. D., and WILLIAMS, R.: Acid-base and metabolic disturbances in fulminant hepatic failure. Gut *16:* 144–149 (1975b).

ROOTH, G. and BANDMAN, U.: Renal response to acid load after phenformin. Br. med. J. *iv:* 256–257 (1973).

ROWELL, L. B.; KRANING, K. K.; EVANS, T. O.; KENNEDY, J. W.; BLACKMON, J. R., and KUSUMI, F.: Splanchnic removal of lactate and pyruvate during prolonged exercise in man. J. appl. Physiol. *21:* 1773–1783 (1966).

SCHRÖDER, R.; GUMPERT, J. R. W.; PLUTH, J. R.; ELTRINGHAM, W. K.; JENNY, M. E., and ZOLLINGER, R. M.: The role of the liver in the development of lactic acidosis in low flow states. Post-grad. med. J. *45:* 566–570 (1969).

SEARLE, G. L. and CAVALIERI, R. R.: Determination of lactate kinetics in the human. Analysis of data from single injection vs. continuous infusion methods. Proc. Soc. exp. Biol. Med. *139:* 1002–1006 (1972).

SOFFER, L. J.; DANTES, D. A., and SOBOTKA, H.: Utilisation of intravenously injected sodium *d*-lactate as a test of hepatic function. Archs intern. Med. *62:* 918–924 (1938).

SRIUSSADAPORN, S. and COHEN, J. N.: Regional lactate metabolism in clinical and experimental shock. Circulation *38:* suppl. 6, p. 187 (1968).

TRANQUADA, R. E.; GRANT, W. J., and PETERSON, C. R.: Lactic acidosis. Archs intern. Med. *117:* 192–202 (1966).

UI, M.: A role of phosphofructokinase in pH-dependent regulation of glycolysis. Biochim. biophys. Acta *124:* 310–322 (1966).

WOODS, H. F. and ALBERTI, K. G. G. M.: Dangers of intravenous fructose. Lancet *ii:* 1354–1357 (1972).

WOODS, H. F.; EGGLESTON, L. V., and KREBS, H. A.: The cause of hepatic accumulation of fructose 1-phosphate on fructose loading. Biochem. J. *119:* 501–510 (1970).

WOODS, H. F. and KREBS, H. A.: Xylitol metabolism in the isolated perfused rat liver. Biochem. J. *134:* 437–443 (1973).

YUDKIN, J. and COHEN, R. D.: The contribution of the kidney to the removal of a lactic acid load under normal and acidotic conditions in the conscious rat. Clin. Sci. mol. Med. *48:* 121–131 (1975).

H. F. WOODS, MD, Department of Pharmacology and Therapeutics, University of Sheffield, *Sheffield S10 2TN* (England)

Lactate in Cerebrospinal Fluid

Lactate in Acute Conditions. Int. Symp., Basel 1978, pp. 115–133 (Karger, Basel 1979)

Cerebrospinal Fluid (CSF) Lactate and Pyruvate in Acute Neurological Situations

J.-P. BERGER and R. FAWER

Department of Medicine, Centre Hospitalier Universitaire Vaudois, Lausanne

Introduction

Summary of CSF Physiology

An approximate quantity of 140 ml of CSF fills the inner cavities of the brain and the subarachnoidal space; to maintain equilibrium of the system in man, this CSF is produced at a mean rate of 0.35 ml/min, and reabsorbed at the same rate [37]. Although it has long been thought that CSF was produced by the choroid plexus alone, it is known today that a significant amount (30% or more) of CSF originates in the brain's extracellular fluid [12, 22, 37, 40]. Therefore, CSF could to some extent be an index of cerebral interstitial fluid [38]. From the inner cavities, CSF flows into the subarachnoidal space, part of it circulating down into the lumbar area and then up into the peri-cerebral subarachnoidal space. CSF then is reabsorbed by the so-called bulk flow mechanism, through arachnoid villi protruding inside the large intracranial venous sinuses [37].

Sources of CSF Lactate

Large amounts of lactate (L), the end product of anaerobic glycolysis, can be produced by nervous tissue (including nervous and glial cells) [6, 16, 47, 49, 50, 52]. Produced within cells, lactate diffuses freely into the extracellular space (ECS) of the brain; as no barrier exists between ECS and the inner cavities of the brain, lactate enriches CSF through the open intercellular channels of the ependymal layer. Lactate reaches the lumbar region by the circulating CSF and diffusion within CSF.

Therefore, in theory, lumbar CSF lactate may to some extent reflect brain (or spinal cord) anoxia severe enough to stimulate tissular anaerobic glycolysis.

Apart from the cells of nervous tissue *per se*, another mode of cellular production should be considered: that of the blood, inflammatory or bacterial cells which are sometimes present in the subarachnoidal space (under pathologic conditions), in cases of intracranial hemorrhage, or meningitis, for example.

Finally, could an increase in blood lactate level have an effect on CSF lactate? Apparently, this effect is minimal, provided the 'barriers' are intact. Indeed, it has been demonstrated that blood-brain and blood-CSF barriers do exist for L-lactate, and that a large arterial blood increase may have almost no effect on CSF lactate [1, 10, 37, 41, 42]. Recent observations [31, 52] suggest that lactate penetrates the barriers slowly, by way of an active transport mechanism. However, in cases where barriers may not be functioning properly, such as in severe systemic acidosis, shock or inflammation at the barrier site, larger amounts of lactate may possibly pass through the barriers.

Clearance of CSF Lactate

Lactate oxidative catabolism takes place within nervous tissue but seems to be minimal [19, 32]. Lactate disappears slowly from the subarachnoidal space [43]; there may be retrodiffusion in the nervous tissue or diffusion through the vessel walls, but CSF bulk flow resorption through arachnoid villi seems to be the main escape route.

These characteristics explain the relative independence of blood and CSF L levels [41], and why CSF L can be a durable indicator of anoxic nervous tissue episodes including those of short duration.

Clinical Interest

Investigators have measured CSF lactate and made correlations with clinical situations for many years now [11, 20, 25, 33, 34]. More recently, a rise in CSF L level has been observed in several pathological conditions: post-traumatic brain edema [7, 14, 29], intracranial hypertension [26, 54], occlusive [18, 35, 46, 53] and hemorrhagic [17, 48] cerebrovascular accidents, post-ictal state [3, 4], meningitis [2, 9, 27, 30], hydrocephalus [45], post-anoxic episodes [36], and some metabolic or toxic encephalopathies [13, 21]. However, the clinician now has to face the problem of interpreting any given result. Since there are many sources of CSF lactate and since elevated CSF L is observed in many diseases, the question arises as to whether this dosage is of any practical interest.

The present work was started in 1972 in the hope of clarifying the following questions: (1) How does CSF lactate behave in acute neurological cases currently entering an emergency ward? (2) Is the CSF lactate value helpful in establishing differential diagnosis and when? (3) Does

CSF lactate have prognostic value, and if so, when? (4) How should a given result be interpreted? (5) Are lactate/pyruvate (L/P) ratio or excess-lactate of more interest than lactate alone?

Material and Method

We measured CSF L and P in patients suspected of having cerebral or meningeal pathology who required a lumbar puncture (LP) in our department from 1972 to 1976. The decision to perform a LP was determined by the clinical situation alone. A LP was only performed in the absence of clinical necessity in cases of brain death.

In all cases reported here, a LP was performed during the first 24 h after admission, and within 36 h of onset of symptoms. CSF glucose, protein level and cytologic examination were performed as well as L and P in every case. A bacteriological evaluation was made whenever infection was suspected.

Methods

From 1972 to 1976, we used enzymatic methods for CSF L and P (Boehringer, Mannheim) [23]. Since then, we have stopped measuring CSF P and have used the Hoffmann-La Roche 'Lactate Analyzer' [44].

Enzymatic Methods

At time of LP, 2 ml of CSF were immediately mixed with 2 ml $HCLO_4$ to precipitate proteins and to prevent any glycolytic reaction. Samples were stored at 4 °C until time of dosage; this technique permits several weeks of storage; nevertheless, dosages were carried out within 10 days.

Lactate Analyzer

Using this technique, native CSF was introduced into the analyzer within 1 h and results were available within 5 min. We tested preservation conditions for CSF samples with this method, as no additive was mixed with CSF to stop potential lactate production *in vitro*; it appeared that lactate did not change at room temperature for at least 2 weeks if cell count was normal, and provided sterility was maintained. If, however, CSF contained cells (blood, leukocytes, tumoral cells or bacteria), lactate increased with time; however, storage is possible for a few hours at 4 °C, in close correlation to the number and type of cells in CSF, but here, the faster the reading, the better.

Comparison between Enzymatic Technique and L Analyzer

The comparison was made with 122 samples from 122 different patients where both methods were used simultaneously in 2 different laboratories totally independent from each other. Figure 1 shows that correlation is good between the two methods.

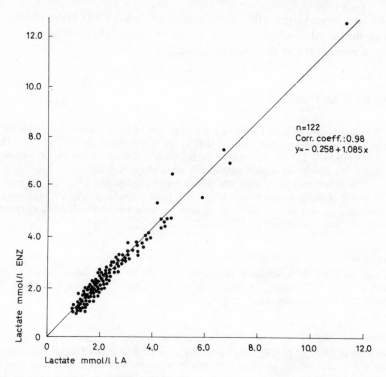

Fig. 1. Values obtained by the enzymatic method (ENZ) are plotted against those obtained by the 'lactate analyzer' (LA).

Control Groups (tables I, III)

CSF studies are limited by ethical considerations: because of discomfort and the slight risk for the patient, LP is not performed without sound clinical justification. The control group consisted of 21 patients suspected of having a cerebral or meningeal pathology which was subsequently excluded.

Comparison with selected papers in the literature shows a good correlation, even if different techniques and different age groups are employed.

Roughly, one may generally consider CSF lactate values below 2.0 mmol/l as being normal (table I).

Mean value for CSF pyruvate was 0.120 ± 0.03 mmol/l in our laboratory; arterial blood lactate was 1.10 ± 0.15 mmol/l and arterial blood pyruvate, 0.10 ± 0.05 mmol/l.

Table I. CSF lactate: normal values

Name of authors	Year	Number of cases	Age	Technique	Mean values (mmol/l) ±SEM
KILLIAN [25]	1925	5	adults	Clausen's	1.66±0.9
MONTANI and PERRET [30]	1964	23	adults	enzymatic	1.60±0.20
POSNER and PLUM [41]	1967	14	adults	Barker/Summerson and enzymatic	1.58±0.03
ZUPPING et al. [53]	1971	?	adults	Barker/Summerson	2.03±0.12
SVENNINGEN and SIESJO [49]	1972	19	3 h to 25 days	enzymatic	1.60±0.32
GERAUD et al. [18]	1973	14	adults	enzymatic	1.81±0.31
BLAND et al. [2]	1974	25	21 days to 12 years	enzymatic	1.58±0.06
PASCU et al. [35]	1974	20	adults	Barker/Summerson (sub.-occip.)	1.79±0.28
BROOKS and ADAMS [3, 4]	1975	13	adults	enzymatic	1.60±0.30
Present series	1977	21	adults	enzymatic	1.80±0.06

Strokes

Several papers in the literature report on CSF L and L/P among stroke cases; results are sometimes conflicting [18, 35, 46, 53], but several authors report increased CSF L in severe cases.

We studied 107 cases of occlusive cerebrovascular accidents; only CSF with normal cell count were selected so as to avoid L production by abnormal cells within CSF; 13 of them presented with subcortical infarcts corresponding to the definition of 'lacunar strokes' [15] and 94 with brain infarction due to a thrombo-embolic mechanism following occlusion of larger arteries. In lacunar strokes (fig. 2, 3), lactate and pyruvate values did not differ significantly from controls.

The brain infarction cases were divided into 4 categories according to clinical evolution: (1) transient ischemic attacks (TIA) (n = 4); (2) minor strokes (with minor sequellae at time of discharge from hospital) (n = 29); (3) medium strokes (left with moderate or severe sequellae) (n = 26); (4) severe strokes (n = 35) either dead (30) or left in 'permanent vegetative state' (5).

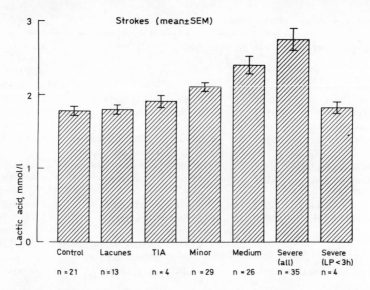

Fig. 2. CSF L at time of admission increases with the severity of the outcome provided that more than approximately 3 h have elapsed since the onset of symptoms. TIA and lacunar strokes do not increase CSF L. The statistical significance between the different groups is given in table II.

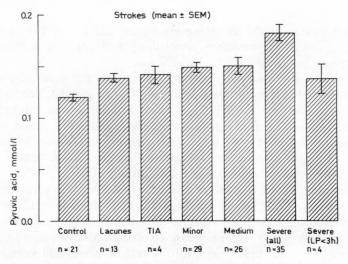

Fig. 3. CSF pyruvate at time of admission in stroke cases.

Table II. CSF lactate in strokes. Statistical comparison between the different categories of strokes (Student's *t*-test)

Control n=21	Lacunes n=13	TIA n=4	Minor n=29 p<	Medium n=26 p<	Severe (>3h) n=31 p<	Severe (<3h) n=4 p<	
—	NS	NS	0.001	0.001	0.001	NS	control
	—	NS	0.005	0.005	0.001	NS	lacunes
		—	NS	NS	0.05	NS	TIA
			—	0.025	0.001	NS	minor
				—	0.025	NS	medium
					—	0.02	severe (>3h)
						—	severe (<3h)

Figure 2 shows that the worse the outcome, the higher the CSF lactate and pyruvate were at time of admission (table II). However, among the severe strokes, 9 cases had only minor CSF L elevation; a more detailed analysis of this group revealed that in 4 cases, the LP had been performed less than 3 h after symptoms occurred; this demonstrated that a certain lapse of time is necessary for lactate to reach the lumbar area. According to our experience, this seems to be between 2 and 3 h. In 2 other cases, death was due to heart attack and pulmonary embolism, but in the remaining 3 cases, no explanation could be found.

It can be concluded from these results that CSF lactate and pyruvate have some prognostic value in stroke cases and that the pattern of 'subcortical infarct with normal CSF lactate and normal EEG' helps differentiate lacunar strokes from the other categories of strokes.

Among a group of 45 patients with infarction in the middle cerebral artery territory (fig. 4), late outcome was plotted against CSF lactate level at admission; again, patients who later died had significantly higher CSF lactate level than the others ($p < 0.02$). However, there was no correlation between CSF L level and the actual neurological deficit. Actual neurological deficit was quantified by a detailed scoring system where zero corresponds to brain death, 215 corresponds to normal nervous system activity.

From figure 4, it is obvious that a severe neurological deficit at admission is the main prognostic factor for late death; however, in some cases low L indicated better prognosis than expected from the neurological examination alone.

Table III. Lactate and pyruvate in CSF. Results and statistical analysis (Student's *t*-test)

	n	Lactate mmol/l±SEM	t-test p<	Pyruvate mmol/l±SEM	t-test p<	L/P ±SEM	t-test p<	Excess lactate	t-test p<
Controls	21	1.80±0.06	0	0.120±0.03	0	15.26±0.66	0	0.00±0.09	0
Lacunes	13	1.82±0.06	NS	0.139±0.04	0.05	13.34±0.57	NS	−0.25±0.09	NS
TIA	4	1.94±0.08	NS	0.142±0.08	0.05	13.76±0.38	NS	−0.18±0.06	NS
Minor strokes	29	2.13±0.12	0.001	0.149±0.005	0.001	14.76±0.70	NS	−0.11±0.07	NS
Medium strokes	26	2.44±0.12	0.001	0.150±0.008	0.005	16.65±0.78	NS	0.18±0.10	NS
Severe strokes	35	2.80±0.15	0.001	0.182±0.008	0.001	15.76±0.85	NS	0.07±0.13	NS
Severe (LP>3 h)	31	2.93±0.16	0.001	0.188±0.008	0.001	16.01±0.94	NS	0.11±0.15	NS
Severe (LP<3 h)	4	1.85±0.08	NS	0.138±0.014	NS	13.87±1.31	NS	−0.21±0.17	NS
Syncope	2	1.76±1.92	–	0.110±0.105	–	–	–	–	–
Post-ictal	10	3.71±0.44	0.001	0.237±0.022	0.001	16.47±1.99	NS	0.16±0.36	NS
Brain death	8	6.20±0.99	0.001	0.222±0.019	0.001	28.66±4.17	0.001	2.87±0.96	0.001
Traumatic LP	3	2.07±0.03	0.02	0.143±0.010	0.05	14.65±1.23	NS	−0.07±0.17	NS
Hemorrhages	22	5.59±0.50	0.001	0.255±0.014	0.001	22.91±2.41	0.005	1.78±0.45	0.001
Hemorrhages (LP<3 h)	1.	2.11	–	0.157	–	–	–	–	–
Viral meningitis	6	1.98±0.11	NS	0.120±0.010	NS	17.05±5.42	NS	0.20±0.13	NS
Bacterial meningitis	8	9.95±1.08	0.001	0.252±0.028	0.001	40.11±5.43	0.001	5.75±1.03	0.001
TBC meningitis	2	8.09±6.92	–	0.144	–	–	–	–	–

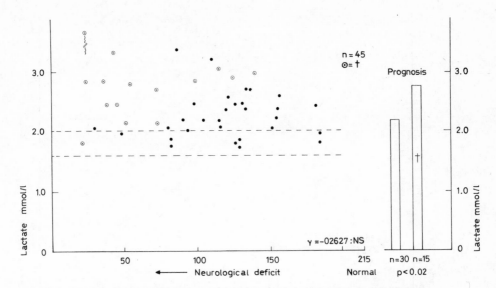

Fig. 4. Comparison between late outcome and CSF L level (right) and actual neurological deficit and CSF L level (left) in 45 patients suffering from infarction of the middle cerebral artery territory (first 36 h). For explanation, see text.

Epileptic Seizures

Generalized epileptic seizure can be considered as severe hypoxia, since ventilation is temporarily impaired. A seizure increases arterial lactate level [39]. Two previous papers report on CSF lactate after epileptic seizures; 10 patients with idiopathic seizures had elevated arterial and CSF L levels. CSF L remained elevated despite a return to normal arterial level in 7 of these patients, from 3 to 6 h after the seizure. Among 13 patients with alcohol withdrawal seizure, CSF and arterial L showed a greater increase [3, 4].

The post-ictal state is characterized by a clear elevation of CSF lactate and pyruvate.

This elevation may be due to systemic L acidosis or to epileptic cortical discharges with higher metabolic requirements when tissular pO_2 is still low; post-ictal hyperventilation with low pCO_2, [5, 8] can play a role although isolated nervous tissue does not show an increase in lactate by decreasing pCO_2 [28].

It is conceivable that all 3 factors play a role. In fact, in experimental animals, it has been shown [8] that curarization and assisted ventilation decrease acid-base perturbations drastically.

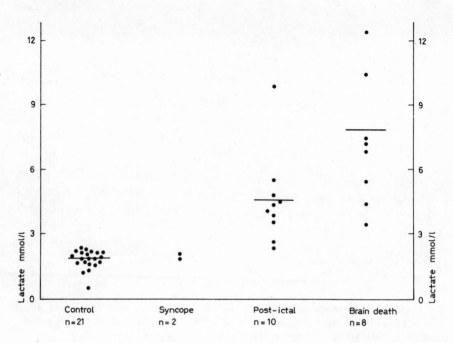

Fig. 5. CSF L level in 2 cases of syncope, 10 cases after epileptic seizure (post-ictal) and 8 cases of brain death compared to the control group.

We followed 10 cases after idiopathic seizure. LP was performed 2–24 h after the fit. In 1 case, CSF L was still elevated (2.4 mmol/l) 4 days after an idiopathic 'grand mal' seizure showing how slow CSF L clearance can be. Figure 5 shows these 10 cases plus 2 cases of syncopal attacks (where history was unclear to begin with).

CSF L can thus help to differentiate between situations which are sometimes difficult to distinguish, if history is not available: syncopal attacks, TIA and post-ictal state; finding an elevated CSF L can be an additional argument in favor of epileptic 'grand mal' seizure. This is valid only for generalized spells, since we have not yet observed the effect of cortical discharges without motor twitching, loss of consciousness, and ventilatory disturbances.

Brain Death

Eight patients presented with all the criteria of brain death; CSF was obtained when diagnosis was confirmed, that is between 30 and 72 h after the causal event had occurred (fig. 3). All cases had 'clear CSF', indicating that tissular anoxia was the main source of L. All cases had elevated values (fig. 5).

General comment on clear CSF. *No patient with CSF L higher than 4.4 mmol/l survived except for 2 epileptic patients. Therefore, elevated CSF L values higher than 4.4 mmol/l, in absence of cells in CSF, can be of vital prognostic importance except in post-ictal states.*

Hemorrhages

CSF may be 'red', because blood has been introduced by the LP itself, or because a hemorrhagic accident has occurred inside the nervous tissue or in the subarachnoidal space; observation of a different red cell count in CSF, at the beginning and end of the procedure, clear CSF after centrifugation, and absence of erythrophages are current methods for differentiating between these two situations.

CSF L measurement is another helpful method, as shown by figure 6: in 3 cases of minor stroke due to brain infarction, CSF was red because of trauma due to LP. Values are clearly lower here than for cases of true (either intracerebral or subarachnoidal) hemorrhages.

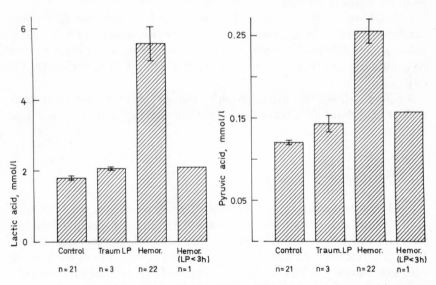

Fig. 6. CSF L and P level in hemorrhages. Results obtained in 22 cases of true hemorrhages (either sub-archnoidal or intracerebral) are compared to 3 cases of traumatic LP and controls (mean±SEM). Differences are highly significant.

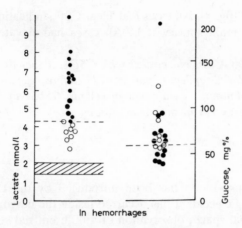

Fig. 7. CSF L level is compared to CSF glucose level in 22 cases of hemorrhages. All patients with CSF L level above 4.3 died (●). A comparable prognostic value is not given by the CSF glucose level.

Thus, the finding of a normal CSF value in presence of blood is a strong argument against a hemorrhage, provided more than 3 h have elapsed since the onset of symptoms.

As shown in the literature [17, 48], CSF L is elevated in hemorrhagic accidents. *Again, among these cases, no patient with a CSF L value above 4.3 mmol/l survived.* This finding may be associated with neurosurgical reports showing bad prognosis when CSF L is elevated [7, 14].

Figure 7 shows that the prognostic value of lactate among those cases is clear, and that CSF glucose level does not give the same information at all.

Meningitis

It has been known since 1924 that bacterial meningitis increases CSF L [33]; several reports confirmed these data, and a paper published in the early 1960s demonstrated that in cases of purulent meningitis, severe metabolic acidosis was associated with higher lactate values contrasting with low values in cases of viral meningitis [30]. Interestingly enough, cases of tuberculous meningitis showed high values, although a lymphocytic reaction was present in the CSF. More recently, the interest of high lactate values among children with partially treated meningitis has also

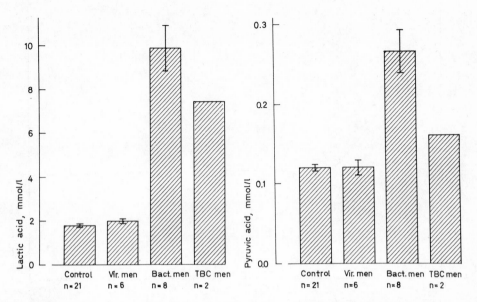

Fig. 8. Comparison between bacterial and viral meningitis with the control group (mean±SEM).

been stressed [9]. More precision in early diagnosis is indeed valuable when one considers the sometimes puzzling differential diagnosis the clinician has to face at the onset of this syndrome [51]. Our data (fig. 8) confirm these observations, showing high lactate values in bacterial infection cases and near normal values in cases of viral infection.

Since this figure was drawn, we have observed two more cases of viral meningitis with higher values: CSF L 3.8 and 4.1 mmol/l, respectively. The first had 6,200 WBC/mm³, 80% polynuclear at the first LP; the second had 4,300 WBC but 95,000 RBC. From this observation, it can be suspected that in viral meningitis CSF L can be more elevated when there is a polynuclear reaction or hemorrhage. However, values remain far below those of bacterial cases.

Among bacterial cases, all survived, even with the highest values measured; *this means that in meningitis, CSF L has no prognostic value, but that CSF L is an aid in differential diagnosis.*

Figure 9 illustrates the evolution of CSF L, glucose and leukocytic reaction in severe meningococcus meningitis; the patient recovered completely. Even after treatment was effective, CSF L remained elevated for several days (i.e. when meningitis was 'partially treated'). The highest arterial lactate level observed was 2.1 mmol/l only.

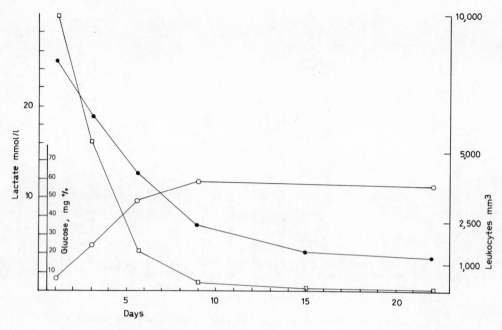

Fig. 9. CSF leukocytes (□), lactate level (●), and glucose level (○) evolution in a case of meningococcus meningitis. For explanation, see text.

L/P Ratio and 'Excess Lactate'

These 2 ratios were thought to provide a better index of cellular hypoxia than does L alone since L production can increase in any situation stimulating aerobic glycolysis [24].

Figures 10 and 11 illustrate that these ratios behave in the same manner, while differing from controls, in only 3 situations: brain death, cerebral and subarachnoidal hemorrhages, and bacterial meningitis.

From these data, it would appear that P measurement is not necessary for the clinician, since L/P ratio and 'excess-lactate' are no more significant than L alone.

Conclusion

Lactate in normocellular CSF is elevated in brain infarcts but not in 'lacunar strokes'. The CSF L level increases in proportion to the severity of the outcome of the stroke. After epileptic seizures, CSF L was elevated

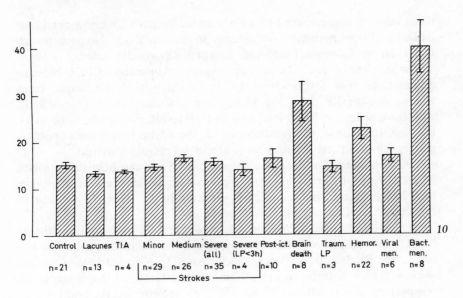

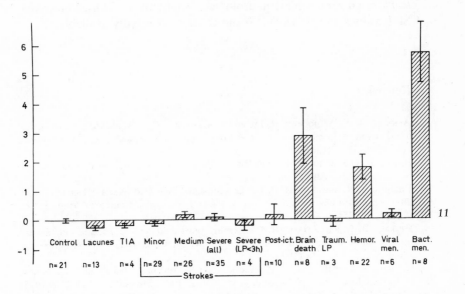

Fig. 10. CSF L/P ratio (mean±SEM) in the different pathological situations studied.

Fig. 11. CSF 'excess lactate' (mean±SEM) in the different pathological situations studied.

when blood L had returned to near normal, so that CSF L is a persistent indicator of brief metabolic disturbance in nervous tissue. No patient with CSF L above 4.4 mmol/l survived, except for post-ictal states.

In presence of 'red' CSF and meningeal inflammation, CSF L helps to differentiate true intracerebral or subarachnoidal hemorrhages from trauma due to LP since the blood L is normally lower than CSF L, whereas true hemorrhages lead to high CSF levels. For results to be valid, the time from onset of symptoms to LP should be longer than approximately 3 h. No patient with CSF L above 4.4 mmol/l survived.

In meningitis, our results confirm that CSF L can help differentiate bacterial (even tuberculous) from viral meningitis, as the L level is very high in the former, but moderately elevated in the latter. In meningitis, the CSF L level has no prognostic value.

In cases of intracerebral pathology where acid-base balance disturbance is not suspected, neither arterial acid-base balance determination, nor L, are needed. If, however, shock, low pO_2, elevated ketone bodies, liver or kidney disturbance are present, then arterial L value is useful for comparison. The results show that CSF L alone provided the best correlation with clinical data. CSF P, L/P or 'excess-lactate' calculations gave no further information. Therefore, CSF P measurement would not seem necessary in current clinical neurological emergencies. Thus, knowledge of CSF L values can be useful to the clinician if rapidly available.

References

1 ALEXANDER, S.; WORKMAN, R. D., and LAMBERTSEN, C. J.: Hyperthermia, lactic acid infusion and the composition of arterial blood and cerebrospinal fluid. Am. J. Physiol. *202:* 1049–1054 (1962).

2 BLAND, R. D.; LISTER, R. C., and RIES, J. P.: Cerebrospinal fluid lactic acid level and pH in meningitis. Am. J. Dis. Child. *128:* 151–156 (1974).

3 BROOKS, B. R. and ADAMS, R. D.: Cerebrospinal fluid acid-base and lactate changes after seizures in unanesthetized man. I. Idiopathic seizures. Neurology, Minneap. *25:* 935–942 (1975).

4 BROOKS, B. R. and ADAMS, R. D.: Cerebrospinal fluid acid-base and lactate changes after seizures in unanesthetized man. II. Alcohol withdrawal seizures. Neurology, Minneap. *25:* 943–948 (1975).

5 BUREAU, M. and BOUVEROT, P. : Blood and CSF acid-base changes and rate of ventilatory acclimatization of awake dogs to 3550 m. Resp. Physiol. *24:* 203–216 (1975).

6 COHEN, E. N.: Reports of scientific meetings. Anesthesiology *36:* 414–416 (1972).

7 GOLD, G.; ENEVOLDSEN, E., and MALMROS, R.: Ventricular fluid lactate pyruvate, bicarbonate and pH in unconscious brain injured patients subjected to controlled ventilation. Acta neurol. scand. *52:* 187–195 (1975).

8   COLLINS, R. C., POSNER, J. B., and PLUM, F.: Cerebral energy metabolism during hyperventilation. Am. J. Physiol. *212*; 864–870 (1967).

9   CONTRONI, G; RODRIGUEZ, C. A.; HICKS, J. M., and ROSS, S.: Cerebrospinal fluid lactate determination: a new parameter for the diagnosis of acute and partially treated meningitis. Chemotherapy *1*: 175–182 (1975).

10 CRONE, C. and SORENSEN, S. C.: The permeability of the blood brain barrier to lactate and pyruvate. Acta physiol. scand. *80*: 47A (1970).

11 SANCTIS, A. G. DE; KILLIAN, J. A., and GARCIA, I.: Lactic acid of spinal fluid in meningitis. Am. J. Dis. Child. *46*: 239–249 (1933).

12 Editor's Note: Cerebrospinal fluid: the lymph of brain. Lancet *Sept.*: 444–445 (1975).

13 EICHENHOLZ, A.; MULHAUSEN, R. O., and REDLEAF, P. S.: Nature of acid-base disturbance in salicylate intoxication. Metabolism *12*: 164–175 (1963).

14 ENEVOLDSEN, E.; COLD, G.; JENSEN, F. T., and MALMROS, R.: Dynamic changes in regional CBF, intraventricular pressure CSF pH and lactate levels during the acute phase of head injury. J. Neurosurg. *44*: 191–214 (1976).

15 FISHER, C. M.: Lacunes, small, deep cerebral infarcts. Neurology, Minneap. *15*: 774–784 (1965).

16 FUJISHIMA, M.; SUGI, T.; MOROTOMI, T., and OMAE, T.: Effects of bilateral carotid artery ligation on brain lactate and pyruvate concentrations, in normotensive and spontaneously hypertensive rats. Stroke *6*: 62–66 (1975).

17 FUJISHIMA, M.; SUGI, T.; CHOKI, J.; YAMAGUCHI, T., and OMAE, T.: Cerebrospinal fluid and arterial lactate, pyruvate and acid-base balance in patients with intracranial hemorrhages. Stroke *6*: 707–714 (1975).

18 GERAUD, J.; RASCOL, A.; BES, A.; GUIRAUD, B.; GERAUD, G.; CHARLET, J. P.; CAUSSANEL, J. P. et DAVID, J.: Equilibre acido-basique et pressions partielles gazeuses du sang et du liquide céphalo-rachidien dans les accidents vasculaires cérébraux aigus. Revue neurol. *129*: 153–172 (1973).

19 GJEDDE, A.; ANDERSSON, J., and EKLOF, B.: Brain uptake of lactate antipyrine, water and ethanol. Acta physiol. scand. *93*: 145–149 (1975).

20 GLASER, J.: The lactic acid content of cerebrospinal fluid. J. biol. Chem. *69*: 539–547 (1926).

21 GUISADO, R. and ARIEFF, A. I.: Neurologic manifestations of diabetic comas: correlation with biochemical alterations in the brain. Metabolism *24*: 665–679 (1975).

22 HAMMOCK, M. K. and MILHORAT, T. H.: The cerebrospinal fluid: current concepts of its formation. Annls clin. Lab. Sci. *6*: 22–26 (1976).

23 HOHORST, H. J.: in BERGMEYER Methoden der enzymatischen Analyse, vol. II, p. 1425 (Verlag Chemie, Mannheim 1970).

24 HUCKABEE, W. E.: Abnormal resting blood lactate. I. The significance of hyperlactatemia in hospitalized patients. Am. J. Med. *30*: 833–839 (1961).

25 KILLIAN, J. A.: Lactic acid of normal and pathological spinal fluid. Proc. Soc. exp. Biol. Med. *23*: 255–257 (1925).

26 KJALLQUIST, A.; SJESJÖ, and ZWETNOW, N.: Effects of increased intracranial pressure on cerebral blood flow and on cerebral venous pO_2, pCO_2, pH, lactate and pyruvate in dogs. Acta physiol. scand. *75*: 267–275 (1969).

27 LAMISSE, F.; GRENIER, B.; CHONTET, P.; ROLLAND, J. C. et GAUTIER, J.: Intérêt du dosage de l'acide lactique dans le liquide rachidien pour le diagnostic des méningites purulentes. Lyon méd. *228*: 591–595 (1972).

28 MATTHIEU, J. M.: Effet des variations de la pCO_2 et de la HCO_3 sur la production de lactate et de la transmission synaptique du ganglion sympathique cervical isolé du rat. Brain Res. *18*: 1–14 (1970).

29 METZEL, F. and ZIMMERMANN, W. E.: Changes of oxygen pressure, acid-base balance metabolites and electrolytes in CSF and blood after cerebral injury. Acta neurochir. *25:* 177–188 (1971).

30 MONTANI. S. et PERRET, C.: Acidose lactique du liquide céphalo-rachidien dans les méningites bactériennes. Schweiz. med. Wschr. *94:* 1552–1557 (1964).

31 NEMOTO, E. M. and SEVERINGHAUS, J. W.: Stereospecific permeability of rat blood-brain barrier to lactic acid. Stroke *5:* 81–84 (1974).

32 NEMOTO, E. M.; HOFF, J. T., and SEVERINGHAUS, J. W.: Lactate uptake and metabolism by brain during hyperlactatemia and hypoglycemia. Stroke *5:* 48–53 (1974).

33 NISHIMURA, K.: The lactic acid content of blood and spinal fluid. Proc. Soc. exp. Biol. Med *22:* 322–324 (1924).

34 OSNATO, M. and KILLIAN, J. A.: Significant chemical changes in the spinal fluid in meningitis. Archs Neurol. Psychiat., Chicago *15:* 738–750 (1926).

35 PASCU, I.; POPOVICIU, L.; PALADE, C.; SIPOS, C.; FARKAS, I., and IOANICS, E. C. S.: Serial lactic acid and pyruvic acid values in cerebrospinal fluid of patients with cerebral hemorrhage and infarction. Rev. roum. méd. *12:* 405–414 (1974).

36 PAULSON, G. W.; LOCKE, G. E., and YOSHON, D.: Cerebral spinal fluid lactic acid follwing circulatory arrest. Stroke *2:* 565–568 (1971).

37 PLUM, F. and SJESJO, B. K.: Recent advances in CSF physiology. Anesthesiology *42:* 708–730 (1975).

38 PLUM, F. and POSNER, J. B.: Blood and cerebrospinal fluid lactate during hyperventilation. Am. J. Physiol. *212:* 864–870 (1967).

39 POLI, S.; KALBERMATTEN, J. P. DE; ENRICO, J. F. et PERRET, C.: Crise épileptique et perturbations acido-basiques. Helv. med. Acta *50:* suppl., 120 (1971).

40 POLLAY, M.: Formation of cerebrospinal fluid. Relation of studies of isolated choroid plexus to the standing gradient hypothesis. J. Neurosurg. *42:* 665–678 (1975).

41 POSNER, J. B. and PLUM, F.: Independance of blood and cerebrospinal fluid lactate. Archs Neurol., Chicago *16:* 492–496 (1967).

42 POSNER, J. B.; SWANSON, A. G., and PLUM, F.: Acid-base balance in cerebrospinal fluid. Archs Neurol., Chicago *12:* 497–496 (1965).

43 PROCKOP, L. D.: Cerebrospinal fluid lactic acid. Neurology, Minneap. *18:* 189–196 (1968).

44 RACINE, P.; KLENK, H. O., and KOCHSIEK, K.: Rapid lactate determination with an electrochemical enzymatic sensor: clinical usability and comparative measurements. Z. klin. Chem. klin. Biochem. *13:* 533–539 (1975).

45 RAISIS, J. E.; KINAT, G. W.; MCGILLIENDDY, J. E., and MILLER, C. A.: Cerebrospinal fluid lactate and lactate/pyruvate ratios in hydrocephalus. J. Neurosurg. *44:* 337–341 (1976).

46 SCHNABERTH, G. und SUMMER, K.: Zum Hyperventilationssyndrom beim schweren zerebralen Insult. Wien. klin. Wschr. *28:* 57–63 (1972).

47 SHANNON, D. C. ; KAZEMI, H.; CROTEAU, N., and PARSONS, E. F.: Cerebral acid-base changes during reduced cranial blood flow. Resp. Physiol. *8:* 385–396 (1970).

48 SUGI, I.; FUJISHIMA, M., and OMAE, T.: Lactate and pyruvate concentrations and acid-base balance of cerebrospinal fluid in experimentally induced intracerebral and subarachnoid hemorrhage in dogs. Stroke *6:* 715–719 (1975).

49 SVENNINGEN, N. W. and SIESJO, B. K.: Cerebrospinal fluid lactate/pyruvate ratio in normal and asphyxiated neonates. Acta paediat. scand. *61:* 117–124 (1972).

50 TAKAYUKI, I.; KATSUHIRO, W.; TAKASHI, K.; KYUHEI, I., and TAKASHI, N.: Lactate in the cerebrospinal fluid and pressure-flow relationships in canine cerebral circulation. Stroke *4:* 207–212 (1973).

51   WHEELER, W. E.: The lumbar tapper's dilemma. J. Pediat. *77:* 747–748 (1970).
52   ZIMMER, R. and LANG, R.: Rates of lactic acid permeation and utilizisation in the isolated dog brain. Am. J. Physiol. *229:* 432–437 (1975).
53   ZUPPING, R.; KAASIK, A. E., and RANDAM, E.: Cerebrospinal fluid metabolic acidosis and brain oxygen supply. Studies in patients with brain infarction. Archs Neurol., Chicago *25:* 33–38 (1971).
54   ZWETNOW, N.: Effects of intracranial hypertension: acid-base changes and lactate changes in CSF and brain tissue. Scand. J. Lab. clin. Invest. *102:* suppl. IIID (1968).

Dr. J.-P. BERGER, PD, Department of Medicine, Centre Hospitalier Universitaire Vaudois, *CH-1011 Lausanne* (Switzerland)

Round Table Discussion

Chairman: C. PERRET

PERRET: Let us begin with a discussion of the first two reports concerning type A lactic acidosis, that is lactic acidosis secondary to tissue anoxia. Has anyone any question or comment?

A listener: Are you entitled to correct severe lactic acidosis with sodium lactate which is an alkalinizing substance?

ALBERTI: If you have a patient who has a high lactate and a severe acidosis, putting in sodium lactate. you are going to depend on lactate metabolism to generate alkali and if he is not generating it in the first place, then it is probably wiser to give bicarbonate if you want to alkalinize him. I do not think there is any single clinical indication in any situation for using lactate in preference to bicarbonate as alkali therapy.

PERRET: Thank you Dr. Alberti. We all agree, I think. I would like to pass on to Dr. Rackwitz who will present some interesting data on lactic acidosis after cardiac arrest.

RACKWITZ: I want to turn your attention to the special point of cardiac arrest. During cardiac massage with ventilation of 100% oxygen, the increase of arterial lactate amounts to about 3 mmol/10 min.

The analysis of the cardiac disturbances shows that the lactate increase in cardiac arrest and ventricular fibrillation is the same. After resuscitation we can divide patients into 3 groups. In uncomplicated rapid recovery, the lactate level will be below 4 mmol/l within 4–6 h. In resuscitation followed by irreversible shock, lactate does not decrease, but shows a further rise, all these patients will die. There is a group of patients in between showing a labile circulation and a delayed normalization of lactate. Blood lactate seems to be, according to these results, a useful parameter for prognosis following resuscitation.

During resuscitation sodium bicarbonate was given. There is a close correlation between lactate and the base excess, as long as bicarbonate is not yet given, but after bicarbonate therapy this correlation is lost. In this situation, the correction of the acid-base status is shown by the measured blood-gas values, but the biological condition is, in my opinion, better characterized by the metabolic parameter of blood lactate. This can be seen in the course of individual cases in whom hemodynamic measurements (besides arterial blood pressure, cardiac index, and pulmonary artery pressures), blood-gas values and urine flow show an equivocal behavior, but blood lactate allows one to differentiate between patients who recover and those who will die after several days. These results are derived from a retrospective analysis. But, provided simultaneous lactate values are available, the following conclusions can already be drawn during the course of treatment. A rapid normalization of blood lactate following resuscitation announces uneventful recovery, even if hemodynamic values are controversial. On the other hand, as long as lactate remains high or even increases, prognosis will be poor or doubtful (cf. RACKWITZ et al.: Intensivmedizin 12/1: 1–23, 1975).

PERRET: Thank you Dr. Rackwitz. Your results are in agreement with those we obtained in the same conditions. As you say, hyperlactatemia per se has no prognostic significance after cardiac arrest, but the evolution of lactate level can give useful information concerning the outcome.

JAHRMÄRKER: I would like to report shortly upon observations on blood lactate in cardiogenic shock and in myocardial infarction in general. Concerning the initial value of blood lactate in acute myocardial infarction upon admission to the hospital, we had the same results as Dr. Perret and others. There is a significant increase of lethality when the lactate is increased upon admission. We prefer this examination upon admission for defining critical cases and enabling us to adjust our monitoring methods accordingly. The maximal values come too late. The prognostic significance applies, as Dr. Perret already mentioned, to cardiogenic shock only and not to other forms of shock. What is the reason for this difference? Shock is the state of insufficient perfusion, as reflected by the increased blood lactate. Etiology and pathogenesis are different, and the prognosis depends on the underlying disease. In cardiogenic shock only there is a correlation between the underlying disease and the poor perfusion in the tissues and in the liver, and in this relatively uniform condition (cardiocirculatory origin and also catecholamines, infusion therapy, etc.), one has a correlation to the prognosis. On the other hand, for instance in septic shock, the actual state of circulation is not indicative, since prognosis depends mainly on the anti-infectious treatment and so on. The next point I want to make is to the meaning of blood lactate during the course of intensive treatment in myocardial infarction. Increased blood lactate, what does this mean during the actual course? For that question we did a retrospective analysis based on 150 patients with acute myocardial infarction. If lactate was never increased we had no lethality at all, and if lactate stayed permanently above 4 mmol/l, lethality was 100%. Other groups are in between. The correlation to hemodynamic measurements, which we did in about $\frac{1}{3}$ of these patients, shows a rough correlation only, but it is just the point I want to make. The lactate shows whether the circulation is or is not sufficient in relation to the needs, but the hemodynamic values do not show if a certain value of the cardiac output is sufficient for a given situation. Similarly, a pulmonary occlusion pressure of 20 mm Hg shows a critical situation and gives therapeutic guidelines, but does not allow one to establish an individual prognosis. Our conclusion from these studies is to follow blood lactate during the course of acute myocardial infarction by lactate determinations within short intervals, about every second hour or in case of changes in the condition of the patient or in therapeutic procedure. The course of lactate until the actual value gives probability values for the further outcome. If we have a lasting increase of lactate to values between 2 and 4 mmol/l or even higher, the patient is judged to be in danger and then we do hemodynamic monitoring, not for prognosis, but to guide the treatment. If we have a permanent increase of lactate above 4 mmol/l despite adequate treatment, there is nothing to be expected from conservative management and we think we should try the assisted circulation already before the shock is completely developed. If balloon pumping is successful, lactate values go down (4 observations) and if not (3 cases) it increases further. In summary, we try to use serial blood lactate measurements as a screening method in acute myocardial infarction to pick out the patients who need hemodynamic monitoring. Also it seems to be possible to detect to an early point of time those who need additional measures such as balloon pumping (cf. JAHRMÄRKER *et al.* in KAINDL 'The first 24 hours in myocardial infarction', p. 51, Verlag Gerhard Witzstrock, Baden–Baden 1977).

ALBERTI: If I may just come in here. I would like to emphasize something that has just been said, and that is that we are in gross danger of overemphasizing the significance of a lactate measurement or even of a series of lactate measurements. We are looking at one index of a whole series of different events. We have the hemodynamic effects on lactate and – something no one has mentioned today, or at least not in any great detail – the effects of catecholamines *per se* in putting up production; we have the effects of steroids putting up lactate production, and various other factors too. We are looking at one isolated metabolite in a sea of disorder and I think we should not try to read too much into the answer.

PERRET: If there are no more comments upon lactate metabolism in shock, I propose to discuss the problems of hyperlactatemia in diabetes.

CONNOR: I think perhaps Prof. Alberti was being a little deliberately controversial when he said that diabetics, *per se*, have no reason to get disturbances of lactate metabolism. I think that the data from Doar and Cramp on lactate utilization in diabetics was fairly reasonable: that they did have impaired lactate utilization. The other thing is that diabetics do get liver disease. Sometimes just a fatty liver, sometimes a cirrhosis. If you have liver disease and diabetes, then you have higher lactate concentrations than if you have liver disease without diabetes and that suggests to me that the diabetes itself is having some effect. Would you like to comment on that?

ALBERTI: The question of liver disease and diabetes is an old and knotty one. Certainly diabetics have a moderately increased incidence of cirrhosis. Certainly cirrhotics have an increased incidence of diabetes and not just the hepatogenous sort. Which comes first is, I think, very difficult to analyze. This was looked at by the Göttingen group in detail some years ago in a very good review and they came to no firm conclusions. As far as the lactate levels go in cirrhotics, with and without diabetes, we certainly have not been able to find a difference when you match them up for degree of disturbance of liver function using a series of different liver function tests. I would be slightly surprised if there were a big difference there.

WOODS: May I make a comment on that? There is a problem here because when you study lactate kinetics in liver disease you are studying one measure of liver function. It is not necessarily valid to say that the absence of abnormalities of classical liver function tests (disturbances in serum enzymes, for example) excludes any liver dysfunction concerning the disposal of a lactate load. In the studies which Henry Connor and others have done, there is virtually no correlation between these two sorts of hepatic function, and you have to separate them out.

ALBERTI: If I may just come back on that. I obviously agree with some of that because liver enzyme measurements, or measurements of enzymes in the circulation are a measure of damage, they are not a measure of function. I think we are all in agreement on that. I was thinking more of function tests, such as albumin which is a synthetic function, such as bilirubin handling, which is a conjugation and transport function, such as the handling of dyes. We were slightly surprised, but in fact found a very close inverse correlation between albumin and mean blood lactate, taken over the day, and now also with glycerol disposal which is an alternative to your lactate disposal and also a positive correlation, again very close, with bilirubin. So the test of damage I think you have got to rule out, but the other tests were correlating remarkably well.

PERRET: I think it is a point well worth making because we are on agreed ground now.

FROSSARD: I would like to ask a question about the treatment of biguanide lactic acidosis. You have said it is important to correct pH. Is there a danger of increasing cerebrospinal fluid acidity in correcting pH in this situation?

ALBERTI: I think the short answer is: yes you will do, and this has been measured in ketoacidosis certainly, where people have put in a lot of bicarbonate and you have got an inverse change in CSF pH taken at the lumbar region. I think Posner and Plum have emphasized that this is not necessarily a change in the cisterna. One of the interesting points is, when you give alkali to a lactic acidotic or to a ketoacidotic, you do decrease respiration, despite this inverse change in lumbar pH. I do not in fact think that is the biggest danger of giving alkali. I do not think we know if that is dangerous or not. I think the possible changes that you see in ... say, oxygen dissociation are a bigger potential problem, although proof is small. A second thing that we have been looking at in acidotic rats is that when we returned pH from 6.7 to 7.25, 7.3 and then looked at the liver and muscle metabolism, we found a big increase in lactate output and the halving of ATP in the liver and a 25% fall in muscle. We are now looking for mechanisms. Against this, I think you still have to say that if you do not put the pH up, you are in dead trouble, literally. The main reasons, I think, being circulatory ones, negative ionotropic effect in the heart, peripheral vasodilatation and so forth; so you end up having to steer a middle course, keeping pH up, but not necessarily shoving it back to 7.4. I think Prof. Cohen has some more data or some comments on that.

COHEN: Just to say that we did actually measure CSF pH on one patient with classical phenformin-induced lactic acidosis. It did fall. I think you have got to get the pH up to normal and the reason why I have said that is purely by a review of the literature in which there is a striking difference in prognosis between those patients who had their arterial pH got up to normal, and kept normal, because it tends to relapse, and in those in whom a rather half-hearted attempt has been made to get it up. Now, if you just put great boluses of hypertonic bicarbonate, or even isotonic bicarbonate, you will run into all these problems with DPG, with CSF pH. So you have got to choose some sort of compromise and again, reviewing those cases that did best in the literature, the pHs were elevated to normal with bicarbonate in a period of time which varied from 2 to 6 h. That seems to me to be the best way of going about it.

PERRET: Thank you Dr. Cohen. We got some experience concerning the evolution of pH after alkaline therapy and maybe Dr. J.-P. Berger could mention it.

J.-P. BERGER: We carefully followed 8 patients who were in severe ketoacidosis, trying to see what was the comparative behavior between arterial pH and lactate and CSF pH. In all cases the mean pH in CSF was 7.21 at the beginning and as we did correct acidosis with bicarbonate 4 h later, the mean CSF pH was about 7.12. Now, what does this mean for the patient? We carefully followed the patients and observed in 4 patients only out of the 8, a mild impairment of consciousness. Even tired patients who had been fighting for hours with hyperventilation were just a bit more sleepy after some hours. So I am not convinced that this decrease of pH has something to do with the behavior of brain. The original idea came from Posner and Plum who reported that correction of acidosis would decrease CSF pH and so determine unconsciousness. In fact, it has not been confirmed. It seems now that osmotic disequilibrium is more important than pH disequilibrium.

COHEN: May I make one last point about this. I think that there is a big difference between the pH correction which is necessary for ketoacidosis on the one hand, and lactic acidosis on the other. I think as we all believe now, you have to do very little apart from rehydrate and give insulin in diabetic ketoacidosis, maybe just a little bit of bicarbonate if the pH is below 7. I would hate to do that to someone with lactic acidosis. I think, although you must not do it too quickly, you have got to make a really good effort to correct the acidosis and perhaps 2–6 h is not a bad compromise.

PERRET: It appears that pH disequilibrium may vary largely for the same degree of acidemia. This probably depends on the rate of development of acidosis, which is itself highly variable from a patient to the other. When CSF pH is not known – which of course is the most frequent situation – it seems preferable to correct acidemia slowly, i.e. in about 3–4 h in order to prevent high gradients of protons between the two compartments. Now, I will ask Dr. J.-P. Berger to present some data collected by Dr. Verdon from our group, which concerns hyperlactatemia in diabetic ketoacidosis.

J.-P. BERGER: François Verdon studied a group of 59 patients with ketoacidosis and he looked for the cause of increased lactate level. On the first slide are reported bicarbonate, lactate, pyruvate and also hydroxybutyrate and aceto-acetate levels on admission. The mean blood lactate level was 2.74 mmol/l. He looked for the 8 patients who had a blood lactate level above 5 mmol/l. There were signs of circulatory failure in only 4 patients. Furthermore, if we consider all patients with initial lactate values above 3 mmol/l, only 8 patients had signs of circulatory failure amongst a total of 24 patients. So circulatory failure in itself is surely not the only factor responsible for hyperlactatemia in this situation. No correlation could be established between blood lactate and insulin dose, the presence or absence of infection, the degree of acidosis and the importance of the respiratory response to the acidosis, the net production of proteins, the severity of dehydration, the presence of hypothermia or, finally, the presence of hepatic failure. But there were rough correlations with mean age, mean blood glucose and mean blood urea level. I do not know if you have any comment?

ALBERTI: That is extremely interesting data. The distribution of your lactates is almost identical with those we found in the smaller group of patients in Oxford from which we derived that figure of about 15% having a value of 5 mmol or more – which is the same as yours. There is one point based on some of Prof. Cohen's work. He has shown a decrease in hepatic blood flow with hydrogen ion and I wonder how much of what we are seeing could be impaired supply of lactate to the liver and whether that could not be a contributory factor. I think we have probably got several factors again. The other possibilities would be whether these patients happened to be ones who had extremely high catecholamines and cortisol levels which we did not look at, but which would again be rather interesting to know. We found this with rats. The lactate rose quite markedly after 5 h bicarbonate infusion when we corrected pH. In fact, lactate went up even more when insulin was administered.

PERRET: How do you explain this increase of lactate with bicarbonate? Does it reflect a stimulation of glycolysis only?

ALBERTI: Well, I think it is mainly that. Our explanation, in the rat so far, is that you have a situation where, for some reason, the ATP has gone down and we have got, for want of a better word, anoxia or anaerobia within the cell. At the same time you open up PFK and you get some glucose metabolism, which is then blocking at the lactate and pyruvate level. That is a very simplistic hypothesis so far.

PERRET: Is Dr. W. Berger here? He has vast experience with phenformin therapy and I would like to have his comments.

W. BERGER: We did a survey in Switzerland last year about the incidence of lactic acidosis and of biguanides used in Switzerland. We have in Switzerland 3 types of biguanide: phenformin, buformin and metformin. We asked all hospitals in Switzerland and we have collected 42 cases of lactic acidosis, 34 under buformin treatment, 6 under phenformin treatment and 2 under metformin treatment. We estimated the relative risk by comparing these numbers with the amount of biguanide consumed in Switzerland. When we calculate this ratio we can say that the relative risk of lactic acidosis in phenformin treatment is 8 times higher than in metformin and 5 times higher in buformin treatment than metformin treatment.

PERRET: Thank you Dr. Berger. Is there any further comment concerning hyperlactatemia in diabetes?

CONNOR: I wonder whether I may make a comment. Prof. Alberti has raised the question of flow limitation of hepatic metabolism of lactate to explain some of the alterations in diabetic coma. Now the problem here is, if lactate was a substance which was completely extracted on one pass through the liver, one would expect there to be obviously a great correlation between hepatic blood flow and lactate removal. So there would be a situation – only a theoretical one – where energy supply in the liver, because of low hepatic blood flow, would be insufficient to provide the 6 mol of ATP required to lift 2 mol of lactate up to glucose. But I do not know of any good data which shows this dependence under normal limits of hepatic blood flow.

ILES: I would just like to comment on this relationship of hepatic blood flow to lactate uptake as I have recently been doing some experiments on this subject in the perfused rat liver. What I have found is that even by dropping the flow by 50%, the lactate uptake is virtually unchanged and the ATP level is the same as with a normal flow, but if we go down to a $\frac{1}{5}$ of normal flow, then the ATP is virtually halved and the lactate uptake is about one third of our normal rate. So far we have only reduced the flow whilst maintaining the oxygen tension constant. We have actually measured oxygen uptake and there is a slight decrease at 50% flow and the oxygen uptake is about, between 1/3 and 1/2 of the normal, even at 1/5 of the normal flow.

ALBERTI: Well, the completely anoxic liver does not clear lactate but, in fact, it can carry out quite a large range of metabolic reactions. It will phosphorylate glucose, it will phosphorylate dihydroxyacetone, it will carry out gluconeogenesis. Interestingly enough you can perfuse a liver in the same way as Dr. Iles and Prof. Cohen do in the presence of cyanide and nitrogen and it will maintain a constant ATP level of about 25% of normal for 2 or 3 h. So ATP is still being generated, it does not run down altogether. Now, under those circumstances, at least one of the limitations which is imposed upon the range of reactions which the liver can do, is that ATP concentration. Whether that ATP concentration is sufficient to satisfy the Km of certain reactions which require ATP as a reactant.

PERRET: We shall have now a comment from Dr. Lambert in Nancy who has some data to present concerning hepatic lesions observed in the course of hyperlactatemia.

LAMBERT: A histopathological study of the liver was carried out systematically in 64 patients, hospitalized in our intensive care unit (Service de réanimation, Centre Hospitalier Regional, Prof. Larcan, Nancy/France), and suffering from pathological hyperlactatemia. These studies were carried out upon hepatic biopsies, made either with Franklin needle under laparoscopy in 10 patients during the decrease of the acute episode or with immediate postmortem laparotomy. The biopsies were immediately fixed and the optical and electronic microscopy studies were carried out according to Petersen's methods, in order to eliminate the alterations due to fixation deficiencies.

From the etiological point of view, the patients were classified into 3 groups:

The first group includes 21 diabetic patients, among whom, 13 with degenerative complications, especially renal, were treated with biguanide, 7 with phenformin and 6 with metformin. In the other 8 patients, an hyperlactatemia was noticed 7 times in the course of a ketoacidosis and once during the decrease of an hyperosmolar coma.

The second group includes 25 patients suffering from various aggressions. In 15 cases, there is a hemorrhagic shock, in 6 cases, a toxi-infectious state and in 9 cases, there are one refractory hypoxemia, 2 severe intoxications with colchicine and dichloropropane, 6 therapeutical accidents and a cardiac liver.

The third group includes severe hepatic affections, 2 fulminant viral hepatites, 12 secondarily aggravated viral hepatites and 4 shocks or severe hypoxia in cirrhotic patients. A metabolic acidosis with a pH inferior to 7.30 and hypocapnia exists in 80% of the first group cases, 72% of the second group cases and only 33% of the third group cases. In all these patients, the arterial lactatemia is superior to 6 mEq/l and the lactate-pyruvate ratio is always superior to 20. The biological signs of hepatic disease are limited to a generally moderate cytolysis and to a hypoprothrombinemia. The optical microscopy study shows constant hepatocellular lesions, but of variable nature, which can be divided into 3 groups:

On one hand, massive steatosis lesions with invasion of the hepatic cells by lipidic micro- and macrovacuoles disseminated in the cytoplasm. That steatosis concerns all the lobules, but there is no change in the lobular architecture.

On the other hand, more or less distinct hepatocellular necrotic lesions, with degeneration of cytoplasm differentiated cell constituents and of nuclei sometimes associated with the rupture of cytoplasm membranes. These necrotic lesions can be localized in centro- and mediolobular areas, without changing the lobular architecture. The hepatic cells of periportal areas are normal. These necrotic lesions can be also massive and can concern the whole lobule, the architecture of which is then completely disordered and unrecognizable.

Finally, the third lesional group includes centro- and mediolobular steatonecrotic lesions. In these cases, although the steatosis and the necrosis are connected, a prevalence of necrotic lesions is generally noticed in centrolobular areas and a prevalence of steatosis lesions in peripheral areas. Besides, in 6 patients, there are lesions of developed cirrhosis, with dissecting fibrosis and regeneration nodules. The distribution of these histopathological lesions changes according to the etiological group with special prevalence of steatosis in diabetic patients, massive necrosis in the course of hepatic diseases and an almost equal distribution of the 3 types of lesions in the group of various aggressions. Moreover, the examination of biopsies colored with orange para-aminosalicylic acid reveals a disappearing of hepatocellular glycogen in all cases, except in five surviving patients, in whom there exists only a decrease of glycogenic stock.

The electronic microscopy study enables one to confirm the lesions observed with optical microscopy and reveals the more or less distinct degeneration of mitochondria. Mitochondrial lesions are of a ballooning type with rarefaction or disappearing of internal crests, clarification of the matrix, even its total disappearance. These ultrastructural changes are noticed clearly at the level of necrotic cells but also at the level of optically normal hepatic cells of perilesional areas and of hepatic cells, the only visible lesion of which, with optical microscopy, is a lipidic overloading.

The study of hepatic biopsies, under the same conditions, in 8 patients, who died in the course of cardiogenic shock and in 4 diabetic patients who died from ketoacidosis did not show similar histopathological or ultrastructural lesion. On the other hand, necrotic lesions are completely similar to those noticed in shock liver, but these lesions appear only late, 2 or 3 days after a long period of shock.

We did not carry out any stereological study which could enable a quantitative comparison of the lesions between the shocked patients with or without hyperlactatemia.

We do not want to discuss the meaning of these hepatic lesions in the pathogenesis of hyperlactatemia, but only to report that they have been constantly observed in the course of a systematical study. These results are obviously an argument in favor of the role of liver disease in the course of hyperlactatemia or of lactic acidosis.

PERRET: Thank you Dr. Lambert. This reminds me of a study that we published several years ago concerning 5 cases of irreversible lactic acidosis, complicating severe centro- or perilobular liver necrosis (Lactic acidosis and liver damage. Helv. med. Acta 35: 977, 1969). None of these patients was in shock. Several arguments led to the hypothesis that lactic acidosis in these conditions was due to an imbalance between a hyperproduction of lactate and a decreased uptake due to the liver damage. Experimental data confirmed this hypothesis. Hyperventilation in dogs increases lactate production which is partly compensated by an increased liver uptake, so that blood level is maintained at a slightly increased level. Clamping of the hepatic artery in these conditions induces a rapidly progressive hyperlactatemia in the absence of shock. Simultaneously, liver uptake ceases and shifts to lactate hyperproduction. Such a vicious circle might play a determinant role in the pathogenesis of so-called 'spontaneous' lactic acidosis.

Now, we could go on to the last chapter which is lactate in cerebrospinal fluid and I suppose there will probably be some questions concerning this point.

KOENIGSHAUSEN: I have a question for Dr. J.-P. Berger. It concerns the diagnosis of cerebral death which is of great clinical importance and should be possible also by determination of the CSF lactate value. I was wondering about your values in your 8 patients. If I remember correctly, you had 4 patients with concentrations between 3 and 6 mmol/l and in my opinion these values are much too low. We observed 5 patients with isolated cerebral death, proved by clinical signs, by EEG and by pathological findings and we had a middle concentration of lactate and CSF of 11.7 mmol/l. The lowest value was 9.7 mmol/l. And these values are nearly the same as those which Paulson found, also with 11.7 mmol/l. He had a larger standard deviation of 4 mmol/l. Are you really sure that your patients had an isolated cerebral death?

J.-P. BERGER: Yes, they all had cerebral death due to cerebral lesions alone, but CSF cell count was lower than 10 WBC/mm^3; this means that no hemorrhage nor meningeal inflammation was present.

I should perhaps make some more comments on this group of patients because lumbar puncture (LP) was not performed during the first 24 h when diagnosis of cerebral death was not certain and when aggravation by possible tentorial herniation due to LP was still possible. LP was performed only when all the criteria of cerebral death were fulfilled and the patients waiting for kidney removal, i.e. 2–3 days after admission; CSF L had thus time to decrease, so I am not surprised that few values are lower than those of Paulson for the causes of death and condition of CSF removal are somewhat different.

KOENIGSHAUSEN: Another question. In which value, in your opinion, is the clearance of lactate out of the CSF and where does the lactate go out of the CSF?

J.-P. BERGER: I would like to have the answer!

KOENIGSHAUSEN: Not in the brain?

J.-P. BERGER: I just mentioned the bulk flow phenomenon for the resorption of the CSF. At least this is one way of escape. Now, are there other ways? In inflammations, one might expect larger escape by increased CSF-blood permeability, but you have seen that in a case of acute meningitis where severe inflammation is present, even in this situation, CSF L level decreases very slowly. So I do not know if there are other significant ways of escape. This explains why CSF L is clinically interesting: after a single anoxic episode of the brain, it remains a longer lasting metabolic indicator of it.

PERRET: Thank you Dr. Berger.

We have arrived at the end of our symposium. A lot of data have been presented during these two sessions and many opinions have been confronted. Nevertheless, several points – some of which are very important – are still not elucidated. This will lead to further studies and, possibly, another meeting in a few years. I am very thankful to all of you who participated in this symposium. I do not wish to close without thanking the Roche Company to have made this meeting not only possible but also most pleasant.

Thank you.

Subject Index